An Introduction to Theatre, Performance and the Cognitive Sciences

Performance and Science: Interdisciplinary Dialogues explores the interactions between science and performance, providing readers with a unique guide to current practices and research in this fast-expanding field. Through shared themes and case studies, the series offers rigorous vocabularies and methods for empirical studies of performance, with each volume involving collaboration between performance scholars, practitioners and scientists. The series encompasses the modalities of performance to include drama, dance and music.

SERIES EDITORS

John Lutterbie, Professor of Theatre Arts and Art
at Stony Brook University, USA
Nicola Shaughnessy, Professor of Performance at the
University of Kent, UK

IN THE SAME SERIES

Affective Performance and Cognitive Science
edited by Nicola Shaughnessy
ISBN 978-1-4081-8398-4

Performance and the Medical Body
edited by Alex Mermikides and Gianna Bouchard
ISBN 978-1-4725-7078-9

Theatre and Cognitive Neuroscience
edited by Clelia Falletti, Gabriele Sofia and Victor Jacono
ISBN 978-1-4725-8478-6

Theatre, Performance and Cognition: Languages, Bodies and Ecologies
edited by Rhonda Blair and Amy Cook
ISBN 978-1-4725-9179-1

Performing Psychologies: Imagination, Creativity and Dramas of the Mind
edited by Nicola Shaughnessy and Philip Barnard
ISBN 978-1-4742-6085-5

*Performing the Remembered Present: The Cognition of Memory in Dance,
Theatre and Music*
edited by Pil Hansen with Bettina Bläsing
ISBN 978-1-4742-8471-4

An Introduction to Theatre, Performance and the Cognitive Sciences

John Lutterbie

methuen | drama

LONDON · NEW YORK · OXFORD · NEW DELHI · SYDNEY

METHUEN DRAMA
Bloomsbury Publishing Plc
50 Bedford Square, London, WC1B 3DP, UK
1385 Broadway, New York, NY 10018, USA

BLOOMSBURY, METHUEN DRAMA and the Methuen
Drama logo are trademarks of Bloomsbury Publishing Plc

First published in Great Britain 2020

Series design by Louise Dugdale
Cover image: *Rain Room* at the Barbican Centre, London © Oli
Scarff/Staff/Getty Images

Bloomsbury Publishing Plc does not have any control over, or
responsibility for, any third-party websites referred to or in this book. All internet
addresses given in this book were correct at the time of going to press.
The author and publisher regret any inconvenience caused if addresses have
changed or sites have ceased to exist, but can accept no responsibility
for any such changes.

A catalogue record for this book is available from the British Library.

A catalog record for this book is available from the Library of Congress.

ISBN: HB: 978-1-4742-5681-0
         PB: 978-1-4742-5704-6
         ePDF: 978-1-4742-5683-4
         eBook: 978-1-4742-5682-7

Series: Performance and Science: Interdisciplinary Dialogues

Typeset by Integra Software Services Pvt. Ltd.
Printed and bound in Great Britain

To find out more about our authors and books visit www.bloomsbury.com
and sign up for our newsletters.

*To my granddaughters, Lily and Eliana Dillon,
and mother, Johanna Lutterbie, with love.*

Contents

List of Figures

Introduction: Intersections

This is an introduction to a fascinating field of theatre and performance studies that brings together cognitive science and the performing arts. In 1959, C. P. Snow, a chemist and novelist, gave a now-famous lecture, "The Two Cultures." Snow saw the seemingly insurmountable divide between the arts and sciences as a hindrance to solving many of the problems facing the world. Half a century later, the sciences and arts are undertaking research to bridge this gap. The questions being asked are profound: What is creativity? What are emotions? What is our relationship with the natural world? With culture? With politics? What is consciousness?

The dichotomy between the two disciplines is, as Snow understood, based on misconceptions. One is that if we learn all there is to know about the brain, we will understand all there is to know about being human. This belief ignores the importance of our lived experience in the development and functioning of our neural networks. Philosopher Shaun Gallagher relates a thought experiment of a brain in a vat: "A disembodied brain, sustained in a chemical bath, seems perfectly capable of experience and cognition as long as the correct neurons are stimulated. On this view, neither body nor environment is required" (Gallagher 2005: 134). He then goes on to consider the types of stimuli needed for the brain to function and concludes that you would need a surrogate body and more:

> Feed-forward or selective mechanisms [that transmit perceptual information to the brain], even if they are specifiable in strict neurophysiological terms, are ultimately the result of interactive relations between the brain, as one important part of the body, and the body as a whole, within a specific environment. We need to consider the brain as part of a holistic system. (143)

To fully understand the brain and how it functions, we need to see it as part of a dynamic system that includes human beings and their complex relations with the world around them. In short, what we need is the knowledge discovered in the arts and humanities.

Another misconception is that emotions get in the way of our ability to reason. We all know that rationality and emotions are diametrically opposed to each other, right? Wrong. Antonio Damasio, a neuroscientist, wrote at the dawn of this millennium a seminal work, *The Feeling of What Happens: Body and Emotion in the Making of Consciousness*. He argues that it is impossible to separate emotion from reason. If all our decisions were purely rational, we would never do anything because the analytical process would never stop. Each question answered gives rise to another question. At some point in our cogitations, the *feeling* arises that the preponderance of the evidence points to one solution being better than another. This may help explain why the best science is sometimes wrong. Scientists know any claim can be disproven by a better analysis, a better set of data ... a better hunch.

The arts are often critiqued by the sciences because they lack the rigor that allows for the advance of knowledge. Scientific advances are of necessity reductionist. In order to prove a hypothesis, each and every variable needs to be controlled to avoid contamination of the data. This is what makes an experiment repeatable. To have merit, an experiment must bring the same results each time it is repeated. Therefore, a scientific claim that cannot be replicated is not acceptable. The arts, on the other hand, revel in complexity. Life is not reducible to what happens in a laboratory. It is messy because of the intricate web of political, cultural, and personal interactions that form the lives we live. The conclusions reached in the arts are equally difficult to prove, perhaps more so because they cannot control all the variables. This does not mean that the knowledge gained is any less important. Splitting the atom is a significant breakthrough, but understanding the art of negotiation is perhaps more important once the power of nuclear fission was unleashed. The arts, and theatre in particular, can provide insights into the dynamics of negotiating everyday life.

Theatre is not constrained by the antiseptic environment of the laboratory. It provides a crucible where we can encounter life while setting aside the cares of the day. Audiences can engage the psychological complexities of relationships, explore the minefield of politics and, as in Michael Frayn's *Copenhagen*, learn about the human side of scientific discovery. Performances are abstract and artificial, but they provide us with the opportunity to see an echo of life in action without the distractions encountered outside the theatre.

C. P. Snow dreamed of a world in which the arts and sciences could escape the silos that define their differences and engage each other in the pursuit of that knowledge, which allows us to live together in health and safety. This book seeks to bridge that divide by exploring ways the disciplines can talk productively. It does not pretend to heal the distrust one side has for the other, but it acknowledges that interdisciplinary conversations are necessary if we are to advance our understanding of how we might live better together.

An Introduction to Theatre, Performance and the Cognitive Sciences brings together three very different fields of study. First, the cognitive sciences include branches of neuroscience, neurobiology, psychology, linguistics, chemistry, physics, and philosophy that focus on issues related to cognition such as consciousness, emotion, memory, gesture, narrative, phenomenology, neural networks, physical therapy, disabilities, brain systems and functions. Second, theatre is fairly self-explanatory and includes the traditional, post-dramatic, immersive, site-specific, applied, and therapeutic, whether in clinics, swimming pools, or prisons. Third, performance is an umbrella term that encompasses theatre and a vast array of cultural genres such as sports, music, performance art, sexuality, gender, the workplace, media, politics, community, and everyday life. Indeed, just about every aspect of life includes some performative element.

This vast array of areas of study is consolidated here into three primary topics: cognitive functions, creativity, and spectatorship. The cognitive functions discussed are dynamic systems, memory, default mode network, plasticity, affordances, metaphor, cognitive integration, neural networks, and 4-E cognition, to give a partial list. These may not be familiar terms, but their relevance should be clear by the end of the book, as will their value to the study of theatre and performance. 4-E cognition requires a preliminary exposition because it is fundamental to the approach taken here and in most cognitive research in the arts and humanities.

The four "E's" are embodied, embedded, enacted, and extended. This quadratic evolved because of the tendency in neurosciences to focus on what happens in the brain, as if that is sufficient to explain the aesthetic process. To counter this assumption, philosophers Evan Thompson, Fransico Varella, and Eleanor Rosch wrote a book entitled *The Embodied Mind* in which they argue that knowledge and consciousness do not exist solely within the confines of the skull but arise from our interactions with the world. They posited, and others have added to their formulation, that the brain is inextricably integrated into and dependent on the body. The brain is embodied. Cognition is part and parcel of all that we do. It is enacted. We gather information from events happening around us and off-load some cognitive activities into the world, such as onto computers and cell phones. It is extended. Finally, the world is our home and we are part of it. Even if we travel to the far reaches of space, we cannot escape our dependence on the environment we live in. It is embedded. While some of these may seem obvious, it is important to be consciously aware of them, as they can easily be taken for granted.

Creativity is perhaps the most mysterious subject that will be discussed in this book, indeed in all cognitive science. No one knows how to explain it. We know it when we see it, and it happens in every field from physics to playing on the football pitch to children at play. The cognitive processes that allow for the emergence of something new in the world are currently unknown. We can say that it is a response to a current situation: we draw on past experience, and the skills we have at our command, and we have an inkling of what it will take to provide a solution. It can happen in the blink of an eye or take months or years to come to fruition. It can happen by accident but more often it requires training to understand how knowledge and imagination can be combined as we engage with and respond to the world around us.

Audiences are necessary for theatre and most performances. Spectators can be present, distant or virtual; they can pay for admission or be your instructor or boss. Like creativity, a performance takes place over time. Whether it is the four-and-a-half hours to see *Einstein on the Beach* or how long you look at a painting, seeing is a temporal experience. Spectating is not only visual, but multimodal, involving hearing, smelling, touching, and tasting. It is not passive but an active engagement with the performance taking place. Audiences can seem complacent, as Peter Handke believed when he wrote *Offending the Audience*, but they are never passive. To take advantage of the active nature of their presence, theatre makers are exploring ways of involving them in the performance through direct participation, by immersing them in a constructed environment, or by leading them through the Meatpacking District of New York City.

It is the goal of this book to understand the intertwining of these three aspects of life: cognition, performance, and spectatorship. You will find very few answers here, but you will discover a new way of thinking about theatre, performance, and perhaps cognitive science as well. Before providing a road map of the trajectory of this book, it might be useful to discuss what it is not.

This is not a science book, and as such you won't find a detailed discussion of the brain, its different areas, or the synaptic cleft. A browser search will provide a wealth of sources where you can delve into these areas of study. There will be some references to areas of the brain, so having ready access to a map of the different neurological areas is a good idea. They are very easy to find given the wonders of the internet. When I began my journey into cognitive science, it was with the synapse. I still find that research incredibly useful, but not mandatory. What you will find is focused more on how the brain functions to make it possible to learn, think, and live in the world.

The scope of the subjects covered here makes it impossible to be equally broad in the kinds and types of theatre and performance covered. This work focuses on Western Theatre, primarily on traditional forms and what Hans-Thies Lehmann calls "post-dramatic theatre." Therefore, there are worlds of theatre that are not touched on or only in passing. Theatre from non-Western countries, including those still colonized, does not find a place, although one is most deserving. It is a question of space (the word limit necessarily set by the editors and a desire to keep the cost of the book as low as possible) and the limitations of my experience with other forms. This is a serious failing, as is the lack of discussions of applied theatre. The work in disability performance, theatre in prisons, and work with communities are immensely important and those engaged in these areas are doing amazing work. The section on further reading will supply works that engage this area more fully.

Finally, this being an introduction to a new area of research, there is limited discussion of post-structuralism, post-modernism, post-colonialism, gender, sexuality, race, ethnicity, and the more recent approaches to integrating these different discourses. The theories presented here are not designed to supplant these important areas of study. Rather, this book should be seen to augment that work by helping us understand why movements like feminism and Black Lives Matter need to happen and suggesting why some approaches to such resistance are more effective. In this way this book engages with politics, as will become evident as you read. Creativity is about change, and while it is important to know who we are, it is equally important to know what we can become.

This is not a book about how to do theatre or make performances. This is a theoretical book that can provide insights into the way theatre is made and how it is received. There is an analysis that most institutions go through on a fairly regular basis. It is called SWOT, and stands for strengths, weaknesses, opportunities, and threats. I focus more on the last two: what impedes creative work, what enhances it, and how we can overcome the former to revel in the latter. I cannot tell you how to do it, but I indicate some ways of thinking and working that will allow you to avoid the traps of simply reproducing what others have done and focus on what you can do. Yet, it does give you a set of tools for evaluating your work, if only indirectly.

An Introduction to Theatre, Performance and the Cognitive Sciences is designed to work by accretion. The chapters discuss specific topics, but the ideas build on each other. What is talked about in the first chapter will help in understanding the second and so forth. I suspect the order you read them doesn't really matter, but reading out of order will not make as much sense as doing it sequentially. As someone who reads in bits and pieces, I cannot complain if you decide to follow my bad example. Perhaps this brief introduction to the chapters will whet your appetite for what is to come.

The book is divided into six chapters with an epilogue. To give you an opportunity to think about the ideas in relation to your own experience, there are tasks that can be done individually or in small groups. They are designed to be simple but not easy. Each will require a degree of introspection and self-reflection. Taking the time (they don't require long periods of navel-gazing) will make what you are reading more meaningful and will help you remember the concepts in greater depth. I hope they will make the material more immediate and more nuanced. Introductory textbooks, whatever the intention of the author, tend to be drier or more jejune than intended. The tasks are designed to enliven your experience and mitigate whatever limitations you discover in my writing.

Landscapes describes the terrain of the book, identifying topographical points of interest along the way. The objective is to introduce terminology you may not be familiar with, such as dynamic systems and affordances. Other terms will be more familiar but may take on unfamiliar resonance once put in relief by cognitive science. With these tools, the journey will begin in earnest. *Culture and the Petri Dish* provides a perspective on theatre and performance by placing them in the wider field of cultural studies. The discussion moves from an aerial perspective on culture to a street level or phenomenological point of view. Memory and plasticity are the primary concepts covered in Chapter 2. One describes how culture becomes ensconced within us, while the other raises the possibility of change from a cognitive perspective. Underlying this discussion is dynamic systems theory, the basis for a discussion on emergence and creativity.

The next two chapters address space and time. *The Worlds of Performance* explores how we encounter the world through the 4-E concepts of embodied, extended, enacted, and embedded. The focus is on how we attend to the world and the affordances that are available to us. This foundation opens a discussion of possible worlds that are created in the theatre and performance from the perspectives of designers, directors, and performers in different venues, spanning the proscenium to immersive forms. Chapter 4 introduces a theory of time based on the work of the French philosopher Gilles Deleuze. It problematizes the sequential logic of past, present, and future by arguing they are always inextricably integrated into the event. With these concepts in mind, we turn to the theatre exploring the importance of time in scenography and text. Productions from Robert Wilson's *Einstein on the Beach* to Théâtre du Soleil's *Les Éphémères* are discussed to help us understand how the use of time affects our experience of the performance's possible worlds.

The next chapter, *The Text*, begins with an exploration of metaphor and cognitive integration. Linguist George Lakoff and philosopher Mark Johnson developed an embodied approach to the origins of metaphor, moving away from a literary definition to an understanding of meaning-making based on physical experience. Their ideas are then complicated by the work of neuroscientist Gilles Fauconnier and philosopher Mark Turner. Their analysis of language focuses on how concepts are combined to make for more nuanced and complex understandings. We then turn to texts. But, in a perhaps unorthodox move, the discussion focuses on the use of silence in the theatre and how the absence of language allows the spectator to engage the text on a deeper level. This decision is also based on the assumption that you already know how to read and analyze texts.

Aesthetics brings the journey to a close by integrating the previous chapters into the heart of theatre and performance: the creation of a complex ecology designed to make a lasting impression on the spectator and the performer. When we go to the theatre, we have a set of expectations based on our history as people and theatregoers. The aesthetic experience is defined as that which exceeds our expectations. It is only when the work of art, be it painting, music, dance, or theatre, challenges our preconceptions that an aesthetic experience takes place. In a very real sense, this is what the book has been about from the outset.

An epilogue has been added to return to the limitations of this book. Instead of just providing an apologia, I discuss work outside the mainstream of theatre and performance. While I love the theatre in all its forms, as long as it exceeds my expectations, I believe the true value of theatre lies in work that is done to improve the lives of others whether it is because of a disability over which they have no control, because of choices they have made that land them in prison, or they have undergone traumatic experiences.

Tim Ingold, in his magnificent book *Lines: A Brief History*, introduces the concept of wayfaring. He differentiates this concept from transport, which he defines as moving from point A to point B. If you concentrate on getting where you are going, he argues, you miss out on a whole lot of life. It is my habit, whenever I visit a new city, to go outside the tourist areas and walk around. I want to see what life is like for the people who live there. I still, after having gone to New York City regularly for the past quarter of century, marvel at the places I have not seen or have not taken the time to see properly. This is a solitary venture that being white and male affords me in most areas. I hope this book encourages you to begin wayfaring individually, or as the Situationist International suggests, in groups of like-minded people with an interest in everyday life. Ingold's writing resonates with the words of Tim Etchells cited in Chapter 3: "What would you like to see on a stage or in a performance space? What would you like to see put in that place that you haven't seen there before? What would make you smile or laugh to see it there? What would scare you?" (Etchells 2008). My hope is that you will all become wayfarers in life and creators of theatre and performances that help people embrace the need for change.

Acknowledgments

A monograph is never produced by one person. There are many who give direct and indirect support to the act of writing. This book is no exception. First thanks go to my co-editor of this series, Nicola Shaughnessy, whose unwavering support has meant more to me than she can imagine. Her scholarship and commitment to working with those with autism and other neurological impairments are exemplary. Mark Dudgeon of Methuen Drama needs to be thanked for his patience during times of great stress that delayed the completion of this manuscript, as does Lara Bateman, and Shanmathi Priya Sampath who helped shepherd this book through the editing process with great care and good humor. Special thanks to Rachel A. Duskewich for her assistance in editing and creating the index. These are necessary and time-consuming parts of the publishing process. To have someone as capable as Rachel is a blessing.

Special thanks go to my fellow musketeers, Amy Cook and Rhonda Blair, who have stood strong with the tools of cognitive science against the winds that have battered our incursions into theatre and performance studies. I would also like to thank Ellen Spolsky, Ben Morgan, Sowon Park, Pat Garrat, Janet Blatter, and the other colleagues who make up Cognitive Futures in the Arts and Humanities, a conference that has energized my work. Bruce McConnachie was of immense help in the early days of my work through the conference at the University of Pittsburgh, the working session in Cognitive Science and Theatre at American Society for Theatre Research (ASTR), and for introducing me to Evan Thompson, whose work is central to mine.

A shout-out also goes to those intrepid colleagues who worked to form the Center for Embodied Cognition, Creativity, and Performance; scientists Nancy Squires, Mary Kritzer, Craig Evinger, Lorna Role, Lisa Muratori, Hoi-Leung Chung, Suparna Rajaram, Richard Gerrig, and Matt Lerner; and philosophers Donn Welton, Bob Crease, and Gabrielle Jackson. In the arts, Meg Schedel, Brooke Belisle, Dan Weymouth, and Judy Lochhead supported the interdisciplinary nature of this work against the reactionary tides of institutional silos. Very special thanks go to E. Ann Kaplan who supported my first forays into the field through the Arts and Sciences Seminar at the Humanities Institute at Stony Brook and has remained a staunch friend and colleague.

It is a pleasure to recognize the students I have worked with, in particular David Rodriguez and Daniel Irving, future stars in the field. I also want to thank the students in the Honors College course taught in Rome in the winter of 2018. They were unaware that they were setting a standard for the tone of this work through their willingness to engage many of the ideas presented here with intelligence and enthusiasm.

I cannot begin to show my appreciation adequately to the lovely Sara Lutterbie, who not only was the first reader of the manuscript, but also created a haven from the pressures of the world, of which there have been many. You are and have always been my soul mate. Thank you. Finally, thanks to my children, Simon and Julia, and to the newest sources of joy and love, my granddaughters Lily and Eliana Dillon.

1

Landscapes

L ooking at a landscape from a fast-moving train, let alone a plane, is to miss the complexity of the vista. It is best seen head-on, while taking time to savor the complexity of what is before you. For anthropologist Tim Ingold, this difference is between transport and wayfaring. The first focuses on getting from point A to point B as efficiently as possible. For the wayfarer, there is a wealth of information to be gleaned as you breathe in the view.

> For the wayfarer whose line goes out for a walk, speed is not an issue. It makes no more sense to ask about the speed of wayfaring than it does to ask about the speed of life. What matters is not how fast one moves, in terms of the ratio of distance to elapsed time, but that this movement should be in phase with, or attuned to, the movements of other phenomena of the inhabited world. (Ingold 2007: 101)

Hiking in the deserts of Tucson, Arizona, the wayfarer is immersed in a world of sun, heat, and sand (see Figure 1). It is a lesson in survival. Looking at art or watching a performance is about wayfaring, being attuned to the world around us. It may not be about survival, but you don't go to the theatre only to see the curtain call. The events on the stage are selected to give you an experience over time. They are highly

Figure 1 *Tucson desert (photograph by author).*

compressed moments that are rich in information that can be seen, heard, smelled, and felt.

Landscape artists, like theatre practitioners, capture only a small slice of all there is to see. Photographs and paintings present a fraction of the world before them; most of the landscape never gets into the final image. It is the same when writing about theatre. Every performance contains an almost infinite number of elements, making it impossible to be all-encompassing. The author, like the landscape artist, selects only those elements most important to communicating his or her vision. Just as artists use their aesthetic judgment in creating a work of art, a scholar has a point of view when looking at the theatre. The perspective I use to explore the landscape of performance is that of the cognitive sciences. For most theatre practitioners, historians, and theorists, this is a very strange approach to the creative arts, but one I hope to show opens new and exciting approaches to understanding performance.

By cognitive sciences I mean those disciplines, empirical and philosophical, that look at how the brain, situated in the body and embedded in the world we experience every day, works and that strive to understand how we act, feel, and learn to be the people we are. The neurosciences, psychology, linguistics, computer science, and philosophy are the primary disciplines that fit this rubric. Each of these is, of course, divided into further areas of research, not all of which are useful for this project. As we move forward with the exploration of theatre and performance, it will become clear which areas of study are most relevant. This is not to say that the others are any less valuable, or that they will not become significant at some future horizon. Rather those cited here reflect how my interests and understanding, of theatre and performance intersect with current approaches to cognition.

Science and theatre are strange bedfellows. Scientific experimentation takes place in the laboratory, with results that are published in disciplinary journals. It is only when there is confidence in the results that the scientist shares them with colleagues. Theatre practitioners present the results of their experiments to the public on opening night confident only that the production is as close to perfection as time allowed. Moreover, science is a reductive discipline, seeking to eliminate every unnecessary variable in order to give the discoveries maximum credibility. Martin Fleischmann and Stanley Pons published research in 1989 claiming that they had succeeded in achieving cold fusion, the ability to generate a nuclear reaction at room temperature (Taubes 1990: 1299). It was subsequently shown that their proclamation was premature because the results could not be reproduced. Unaccounted-for variables skewed the data, rendering conclusions that could not be substantiated. The arts, on the other hand, glory in ambiguity. Art would be uninteresting if all the variables were eliminated from the performance, that is, if every production strove to reproduce exactly the one before it. Theatre in particular is as messy as life itself: complexity is where it finds its power.

Yet, science and theatre have at least one thing in common: the objective of increasing our knowledge of the world. When brought together they can offer new and exciting insights into the realm of performance. I do not pretend to have all the answers, but I hope to use what I have learned from the cognitive sciences to pique your interest in further investigating the amazing world of theatre. For my part, I hope to serve as a guide, a wayfarer who opens doors, inviting you to further exploration. Join me going through the portal of the sciences so that we can begin this voyage into the landscape of theatre.

Performance and theatre

Defining theatre may seem unnecessary. In recent years, however, the approaches to theatre have expanded exponentially. The conventional assumption of the audience sitting passively while watching actors perform a play on stage has been exploded. The proscenium arch has given way to productions that happen in factories, museums, or swimming pools. Theatre is used in interventions for people with autism, Parkinson's disease, and post-traumatic stress disorder. Communities create theatre to deal with divisions of race and class and to educate young people about the dangers of drug and sexual abuse.

In 2012, Connie Hall curated a *site-specific* performance of *Twelfth Night* that was billed as a walking tour of the Meatpacking District of New York City (Anonymous 2010). Audience members were told to meet at various locations around the area where different characters guided them through the performance providing a history of the area as they proceeded. Characters from Shakespeare's play spoke from the windows of a hotel to other characters on the street; Olivia was found on the twelfth floor of a hotel bemoaning her outcast state; and the performance ended on the High Line with cake for all. No audience member had the same experience, and each saw the play only through the eyes of the character that led them through the district.

The British company Punchdrunk has taken over three abandoned warehouses in New York's Chelsea District and dubbed them the McKittrick Hotel, creating an **immersive theatre** event called *Sleep No More*. Audience members are given masks and bundled into an elevator that lets each group off on one of five floors. Each level has a different theme linked very loosely to *Macbeth* with references to Alfred Hitchcock's films *Vertigo* and *Rebecca*. For instance, Lady Macbeth's madness is marked on a floor designed as a lunatic asylum. Actors appear (without masks), briefly enact a scene and then vanish, seemingly into thin air. The audience is left to roam from floor to floor, piecing together a non-existent narrative in what becomes a tantalizingly dream-like experience.

Theatre makers and psychologists at the University of Kent are developing "Imagining Autism," a program that uses improvizational, interactive environments for children with autism. Practitioners engage with the young people, allowing them to determine directions and experiencing for the first time perhaps a validation of their creativity, while learning skills in socialization. At Stony Brook University, students gather true stories from their peers about experiences of alcohol and sexual abuse. These are turned into a fifty-minute documentary performance structured around elements of hip-hop and improvization. The performance is followed by a talkback with health professionals about making positive choices. Petra Kuppers developed "the Salamander Project, an eco-performance project by The Olimpias, a disability culture collective. In the project, disabled people went swimming together and explored themes of stricture and freedom, access and play, biodiversity and border creatures, hanging out together in the wild" (Kruppers 2015: 1). These are just a few examples of **applied theatre**.

This expanded view of theatre raises many questions about the limits of the art form. Is anything that wants to call itself theatre, theatre? For people who prefer a concise definition based on conventional forms, these projects may have theatrical elements but do not qualify as theatre. For those who are excited by the potential of emerging forms, these unconventional approaches are welcome additions to the medium of theatre. For the purposes of this exploration, the more inclusive definition will guide the discussion, acknowledging that this blurs the distinction between theatre and other kinds of performance.

Performance is an even more complicated term that includes music performances, dance, and performance art as well as theatre. These are only the forms that fit the classification of the arts. Richard Schechner, a leading light in the 1960s American Theatre avant-garde, has developed a complex taxonomy of performances that extend to virtually every aspect of life. "Theatre is only one node on a continuum that reaches from ritualization in animal behavior (including humans) through performances in everyday life—greetings, displays of emotion, family scenes and so on—to rites, ceremonies and performances: large-scale theatrical events" (Schechner 1977: 1). So expansive is his definition that it can seem that every aspect of our lives is a performance. There are some who go to this extreme, but then the word loses all significance. A possible, but not completely satisfactory, definition is that performance includes an aspect of *self-conscious awareness*. While this does little to define what performance is, it does identify times when we are not performing, as in non-dream sleep and other unconscious states. Another possible parameter is that performances involve a relationship between self and other. This, too, becomes complicated when asked to determine if looking in a mirror involves a relationship between the self and another.

Task 1

In small groups, challenge each other to define times when there is no performance. Be critical in determining if there is a relationship between self and other in each example, if there is an element of self-conscious awareness of what is taking place. For instance, is there a self/other relationship when I look at myself in a mirror? Is that a performance or not? What does this task tell you about the limits of performance?

As with theatre, I will use an expansive definition of performance while acknowledging that limits exist. Part of the challenge for you is to determine your definition of performance. The purpose of posing these questions—What is theatre? What is performance?—is to encourage you to think about art forms dynamically. Prematurely closing off your understanding of any form can limit your ability to imagine new forms of art. Whether or not you settle on a precise set of elements, there are others who will continue to test the boundaries of performance, exploring new ways of creating artistic experiences.

Theatre as a dynamic system

It is tempting to see theatre as a closed system with actors performing a text for an audience in space and over time precisely in the same way every performance. It doesn't take much to realize that this is too simplistic. For example, actors are influenced by the reactions of the audience, adjusting their performance to responses that differ from night to night. Moreover, both actors and audience can be influenced by elements having nothing to do with the performance, such as what they have had to eat and drink or the temperature of the theatre.

I directed a production of Federico Garcia Lorca's *Blood Wedding*, and in one scene we wanted to see the lovers and their pursuers finding their way through fog. Unfortunately, on certain nights the temperature of the theatre was so cold the fog gravitated to the warmth of the audience, leaving the actors to pretend they couldn't see their way on a mist-less stage. Those who saw the show when the effect worked as planned had a very different and more comfortable experience.

There are also influences that lead spectators to expect a particular kind of performance before they enter the theatre. When Samuel Beckett's *Waiting for Godot* was first performed in Florida, it was promoted as "the laugh sensation of two continents" (Lahr 2009). Needless to say, the audience was not prepared for the existential tragicomedy Beckett had written, putting the performers at a distinct disadvantage with the audience walking out in droves.

To understand theatre and performance, therefore, we need a more complex model that can include, if not necessarily account for, all the variables one encounters in the theatre, or, for that matter, in everyday life. Sophia New and Daniel Belasco Rogers used GPS to precisely track their movement patterns as they carried out routine trips to the store and other places near and far (New and Belasco 2010: 23–31). The variety in the routes they took attests to the inconsistency of their movements, suggesting that even a daily trip to the store involves deviations from the direct route from point A to point B. If there are unexpected occurrences that cause us to go astray when we go to mundane places, we cannot expect any model to account for all the unexpected events that happen during a performance. The best we can do is identify a perspective that can encompass, if not completely explain, those intangible forces that make live performances so exciting.

Fortunately, mathematics provides such a model: *dynamic systems theory (DST)*. To fully understand a dynamic system, we need to differentiate it from its counterpart, the closed system. Our first model of the theatre exemplifies a closed system. Actors perform exactly the same for every audience, and every spectator experiences exactly the same performance as those who saw the show before them and those who will see it afterward. Like turning on a light switch, the same thing happens every time unless the circuit is broken, in which case there is no light, or no performance. In an open or dynamic system, on the other hand, even subtle changes can require an adjustment.

Looking at the conventional theatre experience, we can identify the elements that make up a dynamic system. *Boundary conditions* define the limits in which an event can take place. The proscenium wall creates a separate space for the performers and another for the spectators, who are further separated from the performers by the front of the stage. These are architectural structures that are inflexible, as are the seats in the house and the wings of the stage. *Control parameters* place flexible limitations on what happens. Different plays are performed in the same theatre. Each script identifies different characters and prescribes a set of actions. They are not rigid, however, being open to directorial interpretation, different kinds of casting, etc. Nevertheless, they control what the audience is going to see. Finally, there is the outside force or *perturbation*, such as unanticipated responses of the audience: an unexpected laugh or the absence of an expected one. The performers need to respond to these changes as they occur. Out of these variables, adaptive behavior *emerges* that responds to the needs of the moment, such as covering for another actor who gives you the wrong line. These are the four elements central to understanding a dynamic system: *boundary conditions, control parameters, perturbation,* and **emergence**.

Neuroscientist J. A. Scott Kelso explores the brain as a dynamic system, insisting that it consists of neural networks that interact with each other across brain regions:

> Active, dynamic processes like 'perceiving,' 'attending,' 'remembering,' and 'deciding' that are associated with the word thinking are not restricted to particular brain locations but rather emerge as patterns of interaction among widely distributed neural ensembles and in general between human beings and their worlds. (Kelso 2008: 197)

Most of our lives are spent in what Kelso calls a metastable state: "The metastable regime is … a blend of two tendencies: one for the elements to bind together and the other for the elements to behave independently" (Kelso 2008: 192).

An example of elements that bind together is the ability to ride a bike. As we learn to ride, we learn to use pedals to accelerate and hand levers to brake while maintaining balance and steering at the same time. These skills are bound together, allowing us to say: "Once you know how to ride a bike, you never forget." At the same time, they are used independently of each other. As the road goes up hill, we need to focus on pedaling, while when going downhill, on braking. These elements are both part of and yet independent of the bound skills. In a metastable regime, these elements are coordinated and one rides smoothly. However, when one suddenly rides on loose gravel, causing a skid, the stable state is disrupted by a perturbation that requires an adjustment in behavior to avoid falling. This structure can be applied to virtually any instance that involves a habitual activity, such as getting dressed, using a computer, or playing tennis.

DST can be applied to virtually any event that is open to influences from the environment. Even something as mundane as walking down the street can be analyzed as a dynamic system. DST is used as a meta-theory throughout this book.

Task 2

Observe yourself going down a crowded sidewalk or getting lunch in a crowded cafeteria. Notice when you affect the movements of others and when you need to adapt to the actions of another. What kinds of emergent behavior do you engage in and why?

Task 3

In a small group, identify different kinds of open systems that involve you, such as living with a roommate, or driving a car in a storm. Can you define the boundary conditions and control parameters? Can you remember perturbations or disruptions that required you to adapt your behavior in unexpected ways?

Cognitive structures

I am going to assume we know something about the workings of the brain. For instance, I suspect we all know that neural processes depend on the firing of synapses that transfer an electrical charge from the axon of one cell to the dendrites of another. Moreover, we know the brain consists of different regions, which can be further subdivided by functions. For example, we have a motor cortex that coordinates activity in different parts of the body, such as the ability to wiggle our toes or modulate the activity of our jaws, lips, tongue, larynx, and lungs when we speak. Similarly, we are aware that there are areas associated with perception (visual cortex), abstract thought and memory (prefrontal cortex), reflex actions and voluntary movements (cerebellum), and emotion (limbic system).

How much we need to know about how the brain works is an open question. It may not be particularly useful to know that the motor cortex is in the medial dorsal region of the brain; on the other hand, knowing that the visual cortex is in close proximity to the auditory centers may be useful in understanding the relationship between vision and speech and such metaphors as "I see what you are saying." It is also useful to question popular claims about the brain. The right brain/left brain theory argues that those who are oriented to right hemisphere are more creative, while those who are more adept at using the left are better at computation. While there is some truth to this, it obscures the fact that in most operations we use both sides of the brain more or less equally and that they are in constant communication.

Understanding how our brains are put together and function is a fascinating field, but what is more important for our project, and about which we know less, is how the brain affects behavior, such as remembering past events, making decisions about the future, or, an example we will discuss, tasks as simple as picking up an egg. In executing this action there needs to be coordination between the visual cortex and the motor areas of the brain, and memories of picking up objects of different weights and sizes. This information is necessary for coordinating the neurons that control the muscles of the shoulders, arms, hands, and fingers to avoid dropping or crushing the egg. Some of these neurons activate muscles that allow the hand to grip it, while others inhibit that action, insuring that we use just the right amount of force to make an omelet. This is quite amazing, especially considering that it all happens in milliseconds and without much conscious effort. It should also be noted that we have yet to talk about why we are having an egg, or how we feel about eggs in general, both of which affect the experience.

What this example tells us is that neurons do not work in isolation but cooperate with each other. **Neural networks**, or assemblies of neurons, make such cooperative interactions possible. Neurons associated with the same brain function (vision, movement, etc.) work in tandem with each other, communicating by sharing information (associative interactions) as well as sending data consecutively from one neuron to the next. Cells fire at different frequencies depending on the intensity of the force that "motivates" them. Those that are activated at the same frequency are said to be synchronous with each other. Being in sync with other cells increases their overall effectiveness allowing, in the above example, the egg to be picked up.

Communication between nerve cells not only takes place in bundles, but also between different parts of the brain. David McNeil and Susan Goldin-Meadow at the University of Chicago made a surprising discovery. Looking at the relationship between gesture and speech, they found that gestures are instrumental in processing thought into language. They play a determining role in *what* we are going to say and *how* we are going to say it: "Gesture is not input to speech, nor is speech input to gesture; they occur together" (McNeill 2005: 93). The connection between these seemingly different cognitive processes means that there must be communication between those parts of the brain associated with speech (Broca's areas) and those with motor activity (motor cortex). It seems that our ability to communicate ideas through language and gesture depends on cooperative interaction between these distinct areas.

Theatre consists of intersecting systems that are mutually dependent: acting, lighting cues, stage movement, etc. They rely on each other for the performance to be effective. The relationship between them is not hierarchical, even though there are hierarchies that ensure their proper execution. Systems are at play in everyday performances, too. In this section, we looked at different kinds of systems starting from the internal workings of the brain and moving outward to our interactions with the world in which we live.

French philosopher Gilles Deleuze and psychoanalyst Félix Guattari use the concept of plateaus in their challenging and ultimately rewarding look at the process and limits of change in *A Thousand Plateaus*. They use this concept to move away from hierarchical structures and to envision life as the intertwining of different strands of experience. Existence is seen as the circulation of energy in multiple directions rather than that typified by a top-down relationship, such as when there is a leader and the led. This image of multiple systems interacting in complex ways is useful in talking about relationships between the brain and the environments in which we live.

4-E cognition

The link between physical activity and cognitive processes helps us to understand one of the foundational concepts of this book: **embodied cognition**. Rene Descartes, the seventeenth-century French philosopher, through a radical exploration to discover what could not be doubted, came to the conclusion that the only thing not open to doubt is that he was thinking. This led to his famous statement: "Cogito, ergo sum," "I think, therefore I am." In the deliberation that led to this claim, he argued for a separation of the body and mind, a division that still affects the way we think. The mind is seen as rational and the source of cool intellect, while the body is seen as the reservoir of emotion and irrational heat. This split still permeates our language. We still say, during a heated argument, that the other person should cool down and behave rationally. We want leaders who are cool (rational) in their decision-making.

There are two important ideas to take away from this. One has to do with emotion, and the other with the idea of embodiment. Antonio Damasio, in his 1999 popularization of recent findings in neuroscience *The Feeling of What Happens,* makes a compelling argument that emotion is foundational in decision-making:

> Moreover, the presumed opposition between emotion and reason is no longer accepted without question. For example, work from my laboratory has shown that emotion is integral to the process of reasoning and decision making, for worse and for better. This may sound a bit counterintuitive, at first, but there is evidence to support it. The findings come from the study of several individuals who were entirely rational in the way they ran their lives up to the time when, as a result of neurological damage in specific sites of their brains, they lost a certain class of emotions and, in a momentous parallel development, lost their ability to make rational decisions. (Damasio 2000: 40–1)

Any choice we make requires a belief that one option is better than another. While we would like to believe that we have looked at all sides of the question, the reality is that if we use pure logic we can never come to a decision. Instead, Damasio argues, we feel that one path is better than the others and we take it. Emotion and its corollary feeling play a much more significant role in how we think, even when we focus on making logical choices.

There is a link, therefore, between our physical feelings and our mental processes. Our bodies affect the way we think, whether it is the effect of gesture on what we say or the feeling that one choice is better than another. The phrase we will use to describe this inextricable relationship between the brain and body is **embodied cognition**. Not everyone uses this concept in the same way. Those who study neural behavior believe embodied refers to the activities that take place between neurons, through interactions across synapses or the activity of neurotransmitters. Here, however, we will use the term to signify the inseparability of the brain and the body: the brain cannot be understood without the body and vice versa. They are, in fact, part of a larger system of what it means to be alive. The activities of the body, beyond mere perception, are part and parcel of our cognitive processes. When we are ill for instance, we do not think as clearly as when we are healthy. Stressful situations can give rise to anxiety and depression, severely limiting our ability to see the world clearly. Conversely, when we are well rested and relaxed, we tend to make better decisions.

However, to view the body and the brain as the totality of who we are is to see human beings as closed systems. On the contrary, the environment in which we live constantly affects our behavior. How we behave and how we feel about ourselves is altered by the behavior of other people—whether parents, siblings, roommates, teachers, or a stranger on the street. Similarly, the weather affects our behavior, from influencing what we wear to how we travel across campus or go to the store. Similarly, how we act alters, however minimally, the environment in which we live, much like the proverbial butterfly of Chaos Theory: a flap of its wings on one side of the world causes a hurricane on the other. This interaction between the individual and the world in which they live is called **enacted cognition**: "Enaction is the idea that organisms create their own experience through their actions. Organisms are not passive receivers of input from the environment, but are actors in the environment such that what they experience is shaped by how they act. Many important ideas follow from this premise" (Hutchins 2007: 5). As dynamic systems, we affect the world around us and are affected by it. Indeed, without our perceptions and actions it is hard to imagine what life would be, if we could call it life at all.

This expanded view has led to the development of the **4-E Theory of Cognition**, the four E's being embodied, enacted, extended, and embedded. The first two we have discussed and will return to time and again in the course of the book. The other two are, in some ways, contingent on embodied and enacted. Extended and embedded both relate to our relationship to the environments in which we live. *Extended* recognizes that thought processes are not limited by the boundaries of the brain or the skin, but involve interactions with the world around us. One way to think about this is through Maurice Merleau-Ponty's concept of the *flesh*. In "The Intertwining and the Chiasm," he argues that the flesh is not limited by the material confines of the body; rather it is our ability to perceive the world around us through sight, smell, and hearing (Merleau-Ponty 1968: 152). This encompassing perspective redefines the body by accounting for our perceptual experiences in a way that traditional concepts fail to do. When I drive down the street, changing stoplights or choices made by other drivers affect the decisions I make to ensure my safe arrival at my destination. They bring into that process memories of rules and similar situations, allow me to devise a series of options for action, and question the extent to which I can focus on the music I'm listening to while attending to the flow of traffic. My cognitive processes depend on my ability to extend my attention into the world around me.

Embedded emphasizes that we are always part of the world in which we live, just as journalists were embedded with the troops in Iraq. We are always responding to and acting on the environment, whether it is drinking a cup of coffee, crossing against a "don't walk" sign, performing on stage, or enjoying quiet time in complete solitude. When we indulge in fantasy or the abstract calculations of a math problem, we are nonetheless engaged with the world, shifting our position when we become uncomfortable or disturbed by people walking past or the sudden sound of the air conditioning turning on. We are embedded in the world, and its limits and possibilities determine, to a greater or lesser extent, how we think and what we think about, what we feel, and how we respond.

Task 4

We tend to think of memory being stored in our brains but we use many other devices and strategies for remembering. Make a list of the ways in which you depend on technologies, or other cues in the world around you to supplement memory, for instance by making lists.

Task 5

Discuss in a small group: Imagine a world that is not defined by your experiences with the world in which you live. How successful were you?

In defining three systems—the brain, brain/body, brain/body/world—it may seem that what is being described are Russian nesting dolls, where opening the outer one reveals a smaller doll, in which there is another smaller doll. To think in this way is a mistake, because it suggests that one can be distinguished from the other because we can isolate neurons in the brain, and the brain from the rest of the body, and the body from the world around us. The truth is that we cannot separate them and still continue to live and function as we do. A famous thought experiment asks whether or not a brain could exist in a vat without the body. The logical conclusion is that for it to function as we know it, a mechanical body would need to be built that does all the things our living bodies do: develop perceptual receivers, modulate the ebb and flow of neurotransmitters, and allow for interactive communication with the world in which we live.

The 4-E paradigm is undoubtedly awkward and cumbersome as a formulation. The expansion of what was once simply "embodied cognition" is the result of misunderstandings about what constitutes our involvement with the world and the multiple uses of the original term. As noted earlier, some scientists think of embodiment only in terms of synaptic activity, and others as a means of overcoming the mind/body split. The expansion to four terms is an attempt to be explicit in stating the ways that we are at all times and in all places *inextricably* part of the world in all facets of our existence. For practical purposes, when I use the term *world* I am referring to *4-E cognition*. We now turn to one of the concepts that exemplify how this connection affects our lives.

Affordances

The 4-E's of embodied, embedded, enacted, and extended provide a perspective that necessitates a rethinking of our relationship with the world. If our interactions with the environment are reciprocal, it is because there are qualities in the world that relate to the ways we can interact with it. The physiognomy of birds affords them a different way of interacting with the sky than those of us who are earthbound. The concept developed by James Gibson to define this relationship is **affordances**. "The *affordances* of the environment are what it *offers* the animal, what it *provides* or *furnishes*, either for good or ill … As an affordance of support for a species of animal, however, they have to be measured *relative to the animal*. They are unique for that animal. They are not just abstract physical properties" (Gibson 1986: 127). There are two different ways of thinking about affordances. One is economic—can I afford that computer game—and the other describes how our physiognomy and the architecture of the controller allow us to play the game. What are the limits on how I can use the controller and what does it let me do; what kinds of behavior does it call forth?

Task 6

Think about how you get from where you live to the building in which you take this course. What is it about the environment (physical features or technologies) that makes this movement possible? What physical attributes do you need in order to make the trip?

The design of the control mechanism is based on the opposable thumb. In general, this characteristic allows us to grasp objects, and in the case of the keyboard proves a less central function in writing. However, with the miniaturization of technology, it works better if we hold the remote with our hands and use our thumbs to carry out actions; the same is true with texting on phones. The basic premise is that the organization of the environment permits certain kinds of behavior (you walk up a mountain differently than you walk across a beach) based on how we use our bodies to exist in the world in which we live. Nomadic tribes in the deserts of Sudan interact with their environment differently from the fishermen who live off the shoals of Newfoundland. They have similar physiognomies, but their worlds *afford* them different kinds of behavior.

If you are reading this on a computer or in book form, your relationship with it is different and affords different kinds of action. You can copy a section on a computer and paste it in a new document for the purpose of study. This is a different relationship from note-taking with a book that requires a highlighter, perhaps, and either returning to the book as needed or perhaps typing relevant passages on the computer. Moreover, you will have a different tactile relationship with the text, providing a different kind of experience.

As simplistic as this may sound, learning to climb stairs requires that there are steps to climb. If you never encountered stairs you wouldn't know why they were there, but you would realize, in short order, that they afford you the opportunity of going up. You would know this because you are aware of the potentials of your body and understand the difference between up and down. You could develop a relationship with the stairs that would afford you the opportunity of a new experience, the gaining of new knowledge, as well as getting to your second-floor classroom.

These relationships with the world are taken for granted because they involve either innate abilities or have been learned so well that they have become *second* nature. We do not need to be conscious of what the world affords us because they are part and parcel of our perceptual and intentional interactions. They come to consciousness when, for instance we trip on a stair. Martin Heidegger argues that we become aware of tools when they break down; in other words, when they are no longer to hand, or no longer *afford* us the expected relationship. "When we notice its unhandiness, what is at hand enters the mode of *obtrusiveness*. The more urgently we need what is missing and the more truly it is encountered in its unhandiness, all the more obtrusive does what is at hand become, such that it seems to lose the character of handiness" (Heidegger 1996: 69). Robert Crease and I argue that Heidegger's approach to our experience of affordances is passive and lacks any regard for individual agency (Crease and Lutterbie 2006: 160–79). When we intend to do something, we look for what is needed, for that which affords us the opportunity to be successful. This does not negate Heidegger's stronger claim that when something is not in the circle of our attention, whether because it has not broken down or we do not need it, we are not concerned with its presence. It becomes part of the landscape.

When we experience a breakdown or look for a solution and what is needed is not at hand, the potential for something new—a new way of doing things—emerges. We call on our ability to put information together in new ways and seek a new relationship with the world. This is true not only of interacting with the physical world but also with language as well.

Metaphor, cognitive blending, and gesture

Language remains one of the central mysteries in understanding how the mind works. We can grasp the mechanisms that give rise to speech, locating the areas of the brain where language happens, but until recently there was little understanding of why we put words together in the ways that we do. In the latter half of the twentieth century, language was seen as a system of signs that sought the connection between words and what words mean. Ferdinand de Saussure (1998: 65–7) developed an influential equation:

$$\text{Sign} = \text{Signifier} + \text{Signified}$$

The signifier is the word or symbol used, and the signified is the meaning attributed to it. For instance, in Water = H_2 + O, water is a signifier, as are the symbols for hydrogen and oxygen. They are not the liquid or the gases, but they stand in the place of them; they are signs. If we did not know what they stand for, the equation would be meaningless. Without specialized knowledge the formulas for a "punctured torus," the formula in Figure 2, would be incomprehensible outside of recognizing a few symbols and operations.

$$
\begin{aligned}
SI(\,(a^i b^j a^k b^l)\,) &= (i + K - 2)\,(j + l - 2) + |j - l| - 1 \\
&= (i + K - 1)\,(j + l - 1) - (i + K) + 1 - (j + l) + 1 + 1 + |j - l| + |j - l| - 1 \\
&= (i + K - 1)\,(j + l - 1) - 2\min(i,\,K) - 2\min(j,\,l) + 2 \\
&\le (i + K - 1)\,(j + l - 1) - 2.
\end{aligned}
$$

Figure 2 *Self-intersections number of curves on the punctured torus, Proposition 3.6 (Chas and Phillips 2009).*

Charles Sanders Peirce, working at the same time but independently of de Saussure, realized a fourth element was needed: a referent or interpretant (Peirce 1998: 492–501). In the equation above, the interpretant would be analog filters, electrical engineering, and/or symbolic languages. The interpretant refers to the class of objects, theories, etc., associated with the elements of a sign, providing a context for understanding what is being communicated. While these definitions of signs are useful, they tell us little about why we put words together in the ways we do.

More recently, linguist Georges Lakoff and philosopher Mark Johnson have developed a theory of metaphor that argues that figures of speech have a basis in physical experience. They arise "out of our embodied functioning in the world" (Lakoff and Johnson 1999: 54). When we feel cold, we tend to curl up inside and put on more clothes, and we go outside wearing less when it is warm. The experience of fluctuating temperatures affects our mood and emotional well-being. Similarly, we feel warm when we are embarrassed or blush. These physical effects of temperature form a basis for the ways in which we feel about someone who is sexually appealing. Far from solely linguistic formulations, they claim in *Metaphors We Live By* "*In actuality we feel that no metaphor can ever be comprehended or even adequately represented independently of its experiential basis*" (Lakoff and Johnson 1980: 19; original emphasis). In the theatre, we talk about onstage and offstage, metaphors for visibility, in conjunction with the metaphors of being "in the spotlight" or "in the wings." Similarly, upstage and downstage used to have an experiential interpretant when there were raked stages where performers would literally go uphill the further they went from the audience, and down as they approached the audience. While raked stages still exist, we use the terms even when the stage is flat.

Linguist and cognitive researcher Gilles Fauconnier and philosopher Mark Turner believe that the framework provided by Lakoff and Johnson doesn't go far enough and that a more dynamic approach to language is needed. While Lakoff and Johnson focus on the origins of metaphors, Fauconnier and Turner are interested in how language can be used creatively. Together they developed a theory of **conceptual blending**, also known as **conceptual integration**.

Like metaphor, conceptual blending brings together two dissimilar objects that nonetheless have aspects in common. The **input spaces** are identified with a concept along with their **entailments** or associated qualities (see Figure 4). The **generic space** contains elements that are common to both, in much the same way that common characteristics are depicted in the overlap of domains in a Venn diagram. The **blended space** is what results from the bringing together of the two concepts to create a third. The blend is not static, however, as the blend can continue to "run" as new associations/blends arise from the first.

Our use of language is not only based on past experience, as these theories may imply. Researchers into **gesture** have discovered that how we use our bodies when we speak plays a central role in the organization of thought. There are some gestures that communicate directly, such as pointing (indexical gesture), the peace sign (emblematic gesture), defining a tempo (beat gesture), etc. David McNeill, Susan Goldin-Meadow, and others are more interested in spontaneous movements that are linked to the act of speaking. The

---

Task 7

In small groups or individually, think of other metaphors we use in the theatre, such as "dressing the stage," and think about the physical experiences that give rise to these figures of speech.

usual way of thinking about gestures is that they are used to illustrate what is being said, that they come after language. Experiments reveal, however, that gesturing is simultaneous to speaking or that it may precede language. In one study, they discovered that children learning math show through gesture how a problem is solved before they can express it in words (Goldin-Meadow 2009: 106–11). In other words, our ability to express ourselves linguistically is dependent, in part, on our ability to move. Language is not merely an operation of the brain, but an embodied action.

Meaning-making

When we leave the theatre, we like to feel that we understand what we have just seen. The impact of conventional productions makes this possible by bringing the story to a conclusion that is happy or sad for the primary characters. When Cordelia dies in the arms of King Lear, who then dies of a broken heart and mind, we understand the gravity of his decision to banish his youngest daughter. When Rosalind and Orlando get married at the end of *As You Like It*, we know their love is deeper than the "love at first sight" that ignited the events in the Forest of Arden. Yet the feelings we have after seeing an outstanding performance go beyond a simple interpretation. There are emotions that run deeper than our intelligence can make sense of. Have we "fallen in love" with the actress or actor playing the parts? Was there a beautiful or disturbing line, image, or scene that we can't get out from our minds?

These parts of the experience are central to appreciating the production and help define its aesthetic qualities. In post-structuralist theory, these feelings are what Roland Barthes calls "jouissance" (Barthes 1977: 10–11) and Jacques Derrida calls the "remainder"

(Derrida 2005: 151–2). They are those qualities of a performance that don't easily fit into the overall interpretation of the play, but nonetheless have a lasting effect. I suspect that we need these complex feelings when we leave the theatre to feel fully satisfied by the production. In fact, we may find it boring without them. This is because the show didn't exceed our expectations.

When we go to the theatre we have some *expectations* about what we are going to see. When the audience in Florida went to see *Waiting for Godot*, they were anticipating a fun evening with lots of belly laughs. What they got instead were a couple of tramps, a rock, and a tree. Their expectations were so at odds with what the performance had to offer that they were unable to appreciate Beckett's bleak vision of humanity. The reader-response theorist Hans Robert Jauss calls the point of view of the audience the *horizon of expectations* (Jauss 1982: 88), which define a mindset when encountering any work of art, whether theatre, music, visual art, dance, or even a sporting event. Based on previous experiences, we anticipate what we will encounter when we engage the work of art or hope that our favorite athlete will perform well. The work of art, whether a book, a live performance, or a movie, will present the spectator the *horizon of the text,* or the aspects of the object that define its form, structure, and aesthetic qualities. Most works share some elements, such as characters, actions, and scenic elements. But the artist will give each part of the whole special qualities that evoke emotional and intellectual responses in the spectator.

The performance brings together the two horizons creating the possibility of a truly aesthetic experience. When the horizon of the text meets with the horizon of expectations, there are three possible responses: expectations will be met, fallen short of, or exceeded. When they fail to meet what we anticipate we will most likely be disappointed; those that match our expectations will be satisfying but probably not memorable, giving us only what we thought we were going to get. We might pat ourselves on the back for guessing right, but that won't send us to the moon. What we hope for is an experience that will exceed what we can imagine or, colloquially, "knock our socks off." It is in these moments that we not only gain understanding but also need to respond to the intensity of the experience. Our horizon of expectations will be altered, so the next time we engage with a work of art we will approach the work with different anticipations.

Actors approaching a new role look forward to a challenge. They are going to be disappointed if they find the part is something they can do in their sleep. However, they will respond differently if there is something that piques their interest, and that is intensified if the rehearsal process encourages them to seek solutions outside their comfort zone. In the latter case, their horizon of expectations is exceeded by the horizon of the text.

It would be a mistake to think of either horizon as monolithic, however. Each will have different facets linked to different kinds of experience. A negative critique of many musicals is that you leave the theatre "humming the set." The music is so trite and the spectacle so good that the failure of one area is made up for by the success of another. This is true of all engagements with the horizons. No work of art is so lacking in complexity that there is only one feeling to be

Task 8

Think of the best show you have ever seen, and ask yourself what made it so special. Now think of a show that was boring. What was missing in that show that it failed so miserably? How do these elements affect your horizon of expectations?

gained from it. Some parts of an experience will be mesmerizing, while other parts will be less effective. It is the overall experience that will determine your response to the event.

Plasticity

In *The Graduate*, Mr. McQuire extols the future of plastics as an investment opportunity to Ben. The virtue of plastic, to investors like Mr. McQuire, is its flexibility when heated and its rigidity when cooled. A similar word, elasticity, on the other hand, is flexible at most temperatures but can't hold a shape, returning to its original state. **Plasticity,** the quality of interest here, shares qualities of the plastic and elastic but with significant differences: "*Plasticity* has two basic senses ... the capacity to *receive form* ... and the capacity to *give form*" (Malabou 2008: 5). Modeling clay, for instance, can be sculpted into a particular shape, and then remolded to make something different. It is changeable without losing its fundamental qualities. Neurons have plasticity, and it is this quality that allows us to learn and to "change our minds." The implications of plasticity extend beyond the neural level. "The concept of plasticity has an aesthetic dimension (sculpture, malleability), just as much as an ethical one (solicitude, treatment, help, repair, rescue) and a political one (responsibility in the double movement of the receiving and the giving of form)" (Malabou 2008: 30). We begin, however, with the neural.

A neuron has three main parts: the cell body, dendrite, and axon. The dendrites receive information from other nerve cells and pass the data to the cell body. From the cell body the impulse is sent to the axon, which then passes the signal to other nerve cells. The dendrites can receive the data from a number of different nerve cells, and the signal from the axon can be picked up by a number of neurons. Not all signals are weighted equally, however. The more often a neuron is activated by a particular signal, the more likely it is to pass the information on. The axon will develop "buds" or multiple synaptic clefts that pass the electric charge onto other neurons. Cells can also have numerous points of contact with a neuron, intensifying the charge and thus increasing the likelihood that it will be activated. The larger the number of contact points, the more likely the passage of data will be *facilitated*. As we learn a skill, the more connections will be developed between cells, making it easier to pass information efficiently. However, it also means that it will be more difficult for other neurons to make connections along the facilitated pathway. In terms of behavior, we call these linkages *habits*, that is, something we can do without really having to think about it. These are clearly valuable in terms of species survival. If we had to think about how to run like a toddler learning to walk, we would not do well when pursued by a predator. In these ways, plasticity can become rigid and beneficial.

Not all habits, as I'm sure you know, are good. Bad habits are behaviors that are not sanctioned by the communities in which we live. They are, unfortunately, very difficult to unlearn. To change a way of acting requires that we do two things: *inhibit* neural pathways that support a habit and *excite* connections that strengthen better ways of behaving. The ability to do this, to change, is the value of plasticity.

Over time, theatre practitioners develop techniques that allow them to do their work more efficiently. Professional actors, for instance, need to develop a way of preparing to perform the in three to four weeks allowed for rehearsal. Similarly, directors have to have efficient ways of accomplishing tasks in the timeframes established by unions. Similar things can be said of designers, stage managers, and playwrights.

Problems arise when an actor trained in Stanislavski System is confronted with a play that has other demands. Stanislavski himself faced this problem when he directed Maurice Maeterlinck's *The Blue Bird*. **Emotional memory** did not create the effects demanded by the Symbolist play. The failure of his system to work with this text forced the Russian director to rethink his approach to training actors, leading him to develop an approach based on physical actions, or what is sometimes called **outside-in** acting. Most actors today realize that they need to develop multiple techniques to be versatile enough to respond to the demands of today's theatre. That is, in part, why there are so many different acting studios, each offering another skill set to assist actors in their work, and why plasticity is so important.

Conclusions

This is the landscape of performance viewed through the lens of cognitive science. It is a dynamic field with new forms and demands emerging from the creative energies of practitioners. Theatre makers need to adapt. American Shakespeare performances were considered inferior to the British, when the plays were approached from the perspective of psychological realism. When it was realized that using emotional recall got in the way of communicating the characters' stories, things began to change. Similarly, British acting in realistic plays seemed stilted until the value of character psychology, as espoused in American approaches to actor training, was recognized.

Creativity is about breaking with traditions, confronting habits that come to define the pinnacle of aesthetic excellence for a generation. Innovation arises when the usual approaches are recognized to be ineffective in communicating to an audience with the needed intensity. This is what led to the work of Jerzy Grotowski, Ariane Mnouchkine,

Task 9

Familiarize yourself with any of these theatre practitioners with which you are not familiar. This is a very short list, so in a group add names of men and women who have changed the face of theatre.

Ann Bogart, Joseph Svoboda, Naomi Iizuka, Viola Spolin, Bertolt Brecht, Tadeusz Suzuki, and every other major force in the theatre.

As human beings we are bound by the limitations of our physical being and the values we have been learning from childhood. We are burdened by our history, challenged by limitations of the world we live in, and fueled by our desire to create art. There are parameters defined by boundaries that are implicit in our species and imposed by society. Engaging the world confronts us with limitations and opportunities, affording us new ways of expressing ourselves through new blends of ideas, the opportunity to explore new media and discover new techniques. Just as scientists will never understand the mind completely, the potentials for creating theatre are infinite.

This chapter presents a lexicon that we will use as we explore the world of theatre and performance. What may seem abstract at this point will become more concrete as we look at different aspects of the performing arts. I hope, as we progress through the following chapters, that you will recognize that whatever we say about theatre is equally true in the performance of our everyday lives. We are dynamic beings, with the creative potential to learn new skills that will help us find ever more interesting ways of expressing ourselves, whether in the rehearsal studio, exploring the intricacies of cooking, or engaging with friends. We are not predetermined beings; rather, we are blessed with the potential to change.

2

Culture and the Petri Dish

Peter Brook, in his landmark book *The Empty Space*, asserts that "I can take any empty space and call it a bare stage. A man walks across this empty space whilst someone else is watching him, and this is all that is needed for an act of theatre to be engaged" (Brook 1968: 7). He is right as far as it goes, but as a colleague once protested, there is no such thing as an empty space. Every space has a history that is all too often forgotten. What were the conditions that led to it becoming a performance space? What was it used for previously? Who was displaced? As Hamlet insists, "Come on, you hear this fellow in the cellarage" (*Hamlet*, 1.5). The decision to listen to the ghosts that haunt the space or to ignore them is not only an ethical question—"Those who cannot remember the past are condemned to repeat it" (Santayana 1905: 284)—but also one of memory. What kinds of memory are linked to a space? Or more to the point, what is memory? Are there different kinds? Are there cultural and social memories? If so, how do we become conscious of them? How do they affect performance?

Culture, institutions, and values

A petri dish is used in the laboratory to grow microorganisms. It is a container that holds a medium for the purpose of producing cultures. The virtue of this process as a metaphor is its function of reproducing life, from a few cells to a thriving population. However, if the culture is contaminated by unanticipated variables, the experiment will fail. In the laboratory, this is a problem. In society, it is a virtue.

To think of the environment in which we live as a petri dish allows us to define, in an admittedly simplistic way, the functioning of a social order. Louis Althusser, a French philosopher, wrote an essay in the late 1960s that appears in *Lenin and Philosophy*, "Ideology and Ideological State Apparatuses: Notes toward an Investigation." What he tries to understand is why people hold beliefs and live lives that are contrary to their best interests. He calls these systems of belief *ideology*: "Ideology is the system of ideas and representations which dominate the mind of a man or a social group" (Althusser 1971: 158). By dominate, he doesn't mean that is all we can think about, but that the ideas and representations are so ingrained in our experience that we act on them without thinking. They have become implicit memories.

Psychologists differentiate between *implicit* and *explicit* memory. Implicit refers to things that we know so well that we don't need to think about them. When actors have learned their lines by heart, they depend on implicit memory. They can trust that their words will come to mind when needed. Explicit memories are those things we need to actively remember, such as to bring a script or the time rehearsal begins. Ideology becomes implicit when we act on the "ideas and representations" without being aware we are living by them.

These values and beliefs are taught to us through institutions that Althusser defines as *Ideological State Apparatuses* (ISAs), or institutions that promote belief systems that endorse the authority of the state. Althusser identifies a number of ISAs: the press, religion, political parties, and the family. ISAs are successful when their values are adopted by the members of a political party or religion. "I shall then suggest that ideology 'acts' or 'functions' in such a way that it 'recruits' subjects among the individuals … or 'transforms' the individuals into subjects … by that very precise operation which I have called *interpellation* or hailing, and which can be imagined along the lines of the most commonplace everyday … hailing: 'Hey, you there!'" (Althusser 1971: 174). For example, students enter educational institutions unaware that they are also accepting certain values inherent in the classroom, such as respect for the teacher, obeying commands, and completing assignments on time. These aspects of education have less to do with learning than developing behaviors that will make them good employees and increase the likelihood that they will become productive members of society. In this way, from an early age, students learn to accept that they are part of a hierarchal system based on levels of authority. That is, it becomes "natural" to submit yourself to those in positions of power. To resist these structures results in punishment and/or expulsion from the classroom, diminishing future prospects because students are told that if you are well educated, you have a better chance of succeeding in life.

Althusser asserts that the ideology sponsored by the ISA, like the petri dish, serves to contain the population and maintain order, reducing the need for direct oppression. Greater forms of repression, in most instances, are not needed because the individual internalizes the values promoted by the institution, for instance, believing that submitting to higher authorities is right and proper. ISAs do not only work through regimes of punishment, but also work by actively rewarding their constituents through positive reinforcement: the carrot of economic success, the promise of salvation in the afterlife, and so forth. The latter provide, to continue the metaphor, the nutrients, helping us to grow into socialized members of society, members of a community that upholds similar ideological values.

There are several problems with Althusser's theory. He believes that the representations and beliefs implicit and explicit in the ISAs are part of an all-encompassing ***dominant ideology***, or a unified system of social/political values that everyone adopts. The implication is that all ISAs promote the same system of values, or at least values that aren't contradictory. However, institutions do not share the same ideologies. While degrees of overlap exist, there are areas of contention, if not contradiction, between them, and sometimes internal to them. As a result, people can adopt ideologies from different institutions, causing them to say one thing and behave differently. For example, people are concerned about environmental issues, yet without a thought, purchase technologies made of heavy metals and plastics that become obsolete in three years and endanger the environment. In the United States, people who argue for the sanctity of life support the National Rifle Association despite the rising rate of gun violence. Rather than a homogeneous system, ideology is a patchwork of conflicting value systems that are loosely sewn together. These inconsistencies provide significant opportunities to challenge beliefs and question dominant values in our society.

Task 1

There are more ISAs than identified here. Expand the list, and identify the ideologies (representations, beliefs, and values) that are endorsed by each institution, and how they are revealed in the activities of the organization.

Despite its limitations, there is great value in Althusser's theory. We do learn values through the institutions we embrace without explicitly wanting to. More often than not, our families teach us about love, accepted modes of behavior, and caring for personal needs; at the same time, we are taught respect for authority figures and socially acceptable ways of acting. Much of this happens unconsciously, as if through osmosis. Sociologist Marcel Mauss calls what we learn about cultural and social norms ***habitus***, or "the ways in which from society to society men know how to use their bodies" (Mauss 2006: 78). For example, when children in the United States learn to eat, they hold the fork in the left hand when cutting food. When they prepare to put food in their mouths, they shift it to the right hand. This is very different from European countries where the fork is kept in the left hand throughout. These differ from Asian countries that use chopsticks and some African countries where food is eaten with the fingers. There is no right or wrong way to eat food, but we are brought up in a way that is normal for our culture.

> What takes place is a prestigious imitation. The child, the adult, imitates actions which have succeeded and which he has seen successfully performed by people in whom he has confidence and who have authority over him. The action is imposed from without, from above, even if it is an exclusively biological action, involving his body. The individual borrows the series of movements which constitute it from the action executed in front of him or with him by others. (Mauss 2006: 81)

In gaining habitus, we are gaining markers that provide the foundations of identity. We develop a shared system of forms, values, and beliefs that give us a sense of who we are and our place in the community, and, for better or worse, we learn to differentiate ourselves from others. In positive situations, this provides us with an appreciation for other peoples and cultures, but quite often it creates boundaries of belonging and exclusion and of superiority such as racism and sexism. While we are not all brought up to hold the same ideological positions, we identify ourselves as individuals and as members of a community, or more often multiple communities. Because of these identifications, we have instilled in us values and beliefs about what constitutes right and wrong and our place in society.

I do not wish to imply that this is the only way we learn. Knowledge is accrued through our movement as embodied and active agents. As noted in the first chapter, Lakoff and Johnson contend that we develop metaphors on the basis of physical experiences. The warmth of being held by a caregiver as infants gives heat a positive valence, while being cold is viewed as a negative. These experiences give meaning to phrases such as "the economy is heating up" and "his behavior to her was cold." These values define, in part, who we are and how we interact with others. There are also institutions that actively resist dominant values and beliefs. The Black Lives Matter movement arose in opposition to racism in the States, and the Students for a Democratic Society (SDS) formed to protest US military involvement in Vietnam. Althusser's point is that social institutions have values that we adopt through association, and these beliefs determine to an extent our identity as members of a community. In this way, we become citizens who act based on the relationship people develop in harmony with or in opposition to the morality held by those institutions.

The dynamics of memory

Scientists using petri dishes strive to make a closed system in which the outcomes will be pure strains of the organisms introduced initially. Governments strive to instill in their citizens ideological positions that support the economic system and validate their particular rule of law. But, societies are subject to influences from the outside, giving rise to resistances that range from strengthening a feeling of patriotism to espousing the overthrow of the government. Cultures and societies are open systems subject to external perturbations in the form of immigrations, importation of goods and technologies, or new ways of approaching theatre and performance. There are also internal disruptions due to political or economic corruption, resistance to various biases and stereotypes, and new discoveries that improve our way of life. Some of these will have a very short shelf life, like fads in pop music, while others will have a more lasting impact, such as acts of terrorism. Regardless of derivation of impact, these events leave a residue in the form of cultural memories, memorialized in monuments, museums, and textbooks. These issues and others are ingrained in neural structures of the brain and can bring back joyful experiences, determine political elections, or remind us of horrors that are virtually impossible to forget, as in the case of post-traumatic stress disorder.

Psychologists and cognitive scientists use a variety of names to identify types of memory: long term, short term, working, episodic, implicit, explicit, cultural, historical, social, and I'm sure more. For our purposes, we will identify three types. Two of them derive from French philosopher Gilles Deleuze's discussion of time in *Difference and Repetition*: **pure memory** and **particularities**. The third is **working memory**. Pure memory is more or less the same as long-term memory. It consists of consolidated memories that are not always easy to bring back to consciousness, and as such are only remembered when they are activated, intentionally or by events. Particularities are memories that are relevant to current circumstances and are therefore easily retrievable. They are significant because they help us organize our thoughts about the immediate future. Working memory refers to integrating the particularities with current perceptual experience, or what is happening now.

Psychologist Nelson Cowan explores the storage capacity of short-term memory, and became particularly interested in this question: how many pieces of information can be held in working memory at any one time? Through a number of experiments, he concluded that the magic number is four bits (Cowan 2001: 87–114). This may seem unacceptably small, but it is not as bad as it seems. Each bit of information can actually be a chunk. It is not easy to remember the numbers 1006932948, but if they are "chunked," like 100-693-2948, they are as easy to remember as an American phone number; instead of ten bits of information, there are now three. A chunk can contain a significant amount of information and still be considered one bit.

Cowan uses a spatial metaphor to illustrate his findings. The four bits are contained in the center of attention, or what we are aware of in any one moment (see Figure 3). Outside the circle are the particularities, data that is accessible by consciousness. It can be related

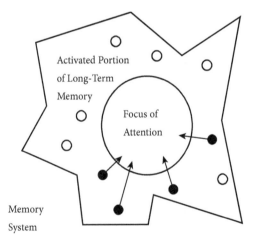

Figure 3 *Attention and the short-term memory system (Cowan 2001: 24).*

to the immediate thought processes directly or by association, or it can be totally unrelated. As I type this, I am aware of the woodstove and occasionally look up from my work to determine whether or not I should put in more wood. Once I am assured the fire is going well, I return to the train of thought that leads to writing these words. Outside of the particularities is the realm of pure memory, or long-term memories that are not relevant to the current situation but may be recalled as needed. As long as the focus remains on the specific event, the bit will remain in the circle of attention. If, however, there is a shift in focus, the items currently in place will be evicted, and more relevant bits of information will take their place.

For example, when performers engage with other actors onstage, their focus needs to be on the immediate present. What is the next line? What is my fellow actor doing to me and how do I need to respond? When am I supposed to move and where? Is the prop in the right place? Is my emotional intensity right for the moment? Fortunately, not all of these are required in an instant, but the actor needs to be ready when the moment comes. These elements and more make up the actor's score ("Creating a score is about making choices and defining a sequence ... Through the manipulation of movement, voice, gesture, memory, and attention, the actor crafts a through line of actions that can be unfolded over the time of the performance night after night. It is a prodigious act of memory" [Lutterbie 2011: 184]) and are available to the performer even if not consciously. As the performance progresses, the correct bits of information will be available as particularities and pulled into the circle of attention as required by the performance. Should this fail, the actor may go up on line, forget blocking, etc.

The information contained in chunks is not solely task driven. There is associative material that will color the situation. For instance, how an actor feels about each of the others on stage will affect the relationship between characters, however marginally. Similarly, in everyday life, our attitudes to persons, places, or things have an impact on our immediate experiences, for better or for worse. Ideology works in the same way. When a director is given a script to read, the style of the theatre she prefers, the kind of play she likes, and who she imagines playing the characters will affect her response to the script. They may have nothing to do with the quality of the play, but they may determine whether or not she decides to direct it and whom she would cast. These biases are as unavoidable as the attitudes people have to immigrants. They will, in all likelihood, be unconscious—few if any people want to be thought a racist—but there are, nonetheless, predictable responses to certain situations and associative memories of past lessons that find their way into the circle of attention. Prejudices are locked in pure memory, seeping out at unexpected moments and in shifting circumstances. They are part of our culture.

Neural plasticity

This can seem to return us to Althusserian negativity: the inescapability of ideology, the impossibility of change. Cognitive science tells us differently. Catherine Malabou, in *What Should We Do with Our Brain*, argues that we have the power to change because of **neural plasticity**. She differentiates between elasticity and plasticity. When the former is stretched and released it returns to its former shape. Plasticity, on the other hand, has "two basic senses: it means at once the capacity to receive form … and the capacity to give form" (Malabou 2008: 5). In the first sense, neural patterns multiply and change through experience, and in the second, our memories give form to our lives and shape our behavior.

Without going into too great detail, neurons consist of the axon, dendrites, and the cell body (Figure 4). This last keeps the cell functioning as a living organism and producing the chemicals needed to pass on synaptic information. The dendrites receive messages from other cells through synapses; the axon passes that data to the

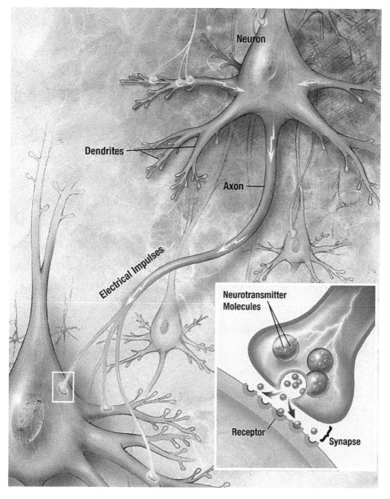

Figure 4 *The synaptic cleft and the body of a neuron (Wikipedia: chemical synapse: https://en.wikipedia.org/wiki/Chemical_synapse).*

dendrites of other cells. Neurons that are frequently activated develop an increased number of "buds" or synaptic clefts on the axon that allow the information to be transmitted to multiple dendrites of other cells. The more frequently a neuron is activated, the more buds it forms, making it easier to pass the information onto other cells. Should it cease to be used as frequently, entropy will set in and the number of buds decreases, making it less likely it will be activated and more likely the memory forgotten.

This does not mean the memory is lost, just more difficult to retrieve. When I first started college, I was fortunate enough to be cast as Edmund in Eugene O'Neill's *Long Day's Journey into Night*, a powerful and formative experience. Since that time, although I had thought about it, I had not read or seen the play until 2003 (some forty years later) when I took my son to see a production with Brian Dennehy, Vanesa Redgrave, Phillip Seymour Hoffman, and Robert Sean Leonard. I was startled to remember what the next line was before it was spoken—not perfectly, but well enough to know that I still remembered large chunks of the script. When teaching at Lon Morris College in East Texas, I directed a play called *Afternoon Tea*. The playwright wanted a certain aria from Bach's *St. Matthews Passion* to be played during the performance. Many years later, Jonathan Miller directed a performance of Bach's work at the Brooklyn Academy of Music. I wondered if I would remember the piece, having not listened to it in the interim. On hearing the first note, I knew. Such was the strength of those memories that they could be recalled.

There is a tendency to believe that as we age, our neural networks become rigid, less plastic, and less able to receive or give form. This is not the case. The adage that you cannot teach an old dog new tricks does not hold. This does not mean we learn as quickly or that we can suddenly change behaviors we have been doing for many years. Habits are persistent: habits of thought as well as behavior. Throughout our lives we are learning. Tim Ingold, in *The Life of Lines*, argues this point from an anthropological point of view. "Since at every moment, the human must resolve not what he is but what he is going to be, at no point can the process arrive at a final conclusion. Fulfillment is ever-deferred, ever 'not yet.' Humans, wherever and however they live, are always humaning [*sic*], creating themselves as they go along" (Ingold 2015: 140). Much of what we learn in creating ourselves is a repetition of what has gone before, but as Deleuze argues, no repetition is ever the same: it is always with a difference. Each remembered experience gives form, however minimally, to the neural networks of memory, and by extension to who we are, our identity. It is in this sense that we are creating ourselves, perhaps discovering things about ourselves that we didn't know before.

We are building on a foundation of what has been, on the memory of times past. But, we continue "creating ourselves." Some neural patterns are resistant to change and focus on keeping things as they are or imagine them to be. In repeating habitual practices, they make the synaptic connection stronger, generating new buds and developing a greater sense of security. Others who seek new experiences are less risk averse. But these too are patterns, habitual ways of being that are strengthened with each foray into the unknown. It requires a powerful, possibly traumatic experience to change a deep-seated habitus.

Task 2

Ask yourself whether you prefer things as they are or if you seek out new experiences. Realize that no one is completely one or the other. Make a list of patterns of behavior that define your identity, not shying away from difficult issues like sexuality, ethnicity, and moral values. Think about what it would take for you to break the pattern, even if it is not something you are likely to do.

Technique

There is another aspect of memory that is especially important in theatre and performance: that is *technique*. Philosopher Hubert Dreyfus developed a theory of mastery (Dreyfus 1997: 17–28), identifying the five stages the individual goes through before claiming expertise. Much like an apprentice, the road to becoming a master involves moving from novice through advanced beginner, competence, and proficiency to expertise. Actors, for instance, generally begin learning the trade through a series of improvizations designed to awaken themselves to their creative potential. When the teacher feels the students learned these exercises, text is introduced and students learn to read a text for the purpose of performance. At this time, courses in voice and movement are often introduced, helping the young actors develop their physical instrument. Scene work may be followed by looking at different styles of plays from classics to the absurd, along with stage combat, mask work, etc.

The hope is that by the time the students leave the studio they will have a basic proficiency, allowing them to enter the profession. In most cases, expertise will not yet be achieved. Work that puts what they have learned into practice in productions is necessary. "The expert not only knows what needs to be achieved, based on mature and practical situational discrimination, but also knows how to achieve the goal" (Dreyfus 1997: 22). It is true for me, and I expect many teachers, that you don't really learn something until you teach it. The same is true with performing. Only by practicing skills in different settings will they become reliable in their use, and actors will know when to use which tool effectively and efficiently. It is not a solitary activity. Most actors acknowledge that they learn from talking and working with other actors about their process and different skills to solve specific problems. Mastery is only achieved when the performer has an approach to acting that is sufficiently flexible to take on most any role, having the techniques to solve the problem of playing King Lear or Lady Macbeth.

Becoming an expert in any field involves a prodigious feat of memory. An adept improviser does not need to think about the first rule of improvization: always say "yes." They know how to pay attention to what they are given without effort and how to respond in a way that keeps the improv alive. All these basic necessities for success have become ingrained and are now implicit memories. They can be brought into play as needed without invitation. A young Robin Williams was interviewed by Dick Cavett on *The Dick Cavett Show* in 1979. Toward the end of the show, Cavett and Williams involved the audience in an extended improvization that moved across a wide range of topics. The transitions from one to the next were seamless: each relationship was fresh, unexpected, and hilarious. There are many examples in the history of performance from *commedia dell'arte* to Monty Python, where the creative impulse is expressed with the full support of the lessons learned over a lifetime of work. But, because they are experts, they give the appearance of effortlessness and pure spontaneity.

Expertise is perhaps most clearly on display in sports. A football player (soccer in the United States) does not have time to think about where the ball is, how to use his feet to move it in the right direction, or whether or not to feint. If he needs to think about it, he has already lost the ball. He needs to decide in a fraction of a second whether to hold the ball, pass it, or take a shot on goal, whether to push forward or hold back. Similarly, the defender doesn't have time to decide what his opponent is doing; he needs to anticipate, based on how the other is moving, where the ball is going to go and when. The player needs these skills to be deeply ingrained, and yet be flexible enough to improvize on the spot.

Actors' techniques need to be equally planted in their neural networks. The major difference is that while sports are competitive, theatre is cooperative. Performers need to support the work of the other actors on stage and be open to variations in their performances to support their discoveries. While these qualities are aspects of technique, they also form a culture that sets expectations about what constitutes good theatre and what is of value in the art. Lee Strasberg formed the Actors Studio based on Stanislavski's concept of emotional recall. The Studio promoted a style of acting based on internalized investigations of emotion. The success of his work established a benchmark for good acting in the United States, despite its inappropriateness for classical theatre, such as Shakespeare or Moliere. His legacy still holds sway in the States, not only determining how people should act, but also setting expectations that new plays will accommodate this style of acting. Only with the influx of European (Lecoq and Barba) and Asian (Suzuki) acting styles has this hold on American theatre begun to diminish. Through the change in neural synapses and communicating these changing beliefs about what constitutes good acting and how acting should be taught, the culture is changing.

Cultural capital

Societies tend to be hierarchical. Certain individuals or groups of people are seen to be more important and powerful. Scales of difference are generally based on political and economic power, giving rise to classes defined by success in tandem with bias based on skin color and/or ethnic background. These distinctions explain numerous examples of discrimination, both positive and negative. What are not clarified by this structure are feelings of superiority among populations in the same class, that is, people of similar backgrounds, wealth, and authority. The French philosopher Pierre Bourdieu addresses this issue in his 1979 book *Distinction: A Social Critique of the Judgement of Taste.*

Bourdieu here develops the concept of **cultural capital** (Bourdieu 1987: 53–4). When we think about capital, we generally think about money or other monetary instruments and assets that can be used to generate wealth. Bourdieu is more interested in the ways people differentiate themselves from each other based on taste, or the ability to evaluate objects and ideas using socially defined scales of excellence. Wine connoisseurs are identified by their ability to rate wines based on color, viscosity, aroma, and flavor. They are not inherently better than those who only know what kind of wines they like, but through learning about different varietals, regions, micro-climates, and vintages, they set the standards for defining great wines. The 2012 documentary film *Somm* follows four men who are trying to become Master Sommeliers, the highest rank achievable for the tasting of wines. Those who eventually achieve this honor have spent years gaining the knowledge and experience to be judged experts in the field. Their status is not based on wealth or power, but on the education of the senses and through memorization of facts and sensual experiences with wine.

People of the same economic class create hierarchies among themselves using as qualifiers education and competence in areas of culture. These measures allow for distinctions to be made between those who are culturally "literate" and those who are less so. "The more legitimate a given area, the more necessary and 'profitable' it is to be competent in it, and the more damaging and 'costly' to be incompetent" (Bourdieu 1984: 86). Legitimacy is the ability, for instance, to distinguish good from bad or indifferent art. The snobbery and arrogance often attributed to academics are based on their education and their tendency to assume that their knowledge makes them better able to appreciate art and ideas because their years of learning have provided them with taste. "Taste is first and foremost distaste, disgust and visceral intolerance of the taste of others" (Bourdieu 1984: 56). Hedda Gabler, in Ibsen's play of that name, looks down on her Aunt Julie's poor taste, Mrs. Elvsted's failed marriage, and Ejlert Lövborg's lack of moral fiber. Her desire to be free from social constraints and to control the lives of "lesser people" leads her to destroy Lövborg and ultimately end her own life.

The assumption of values that create classes within classes leads Althusser to assert that the individual is an ISA, adopting an ideology that privileges her identity over that of others. In so doing, the hierarchies that support the political and economic status quo are reproduced and promulgated in constructing relationships with others. Such divisions make organizing for political and cultural changes more difficult. Celebrity actors frequently get special privileges, such as their own dressing room, creating class divisions and illusory states of superiority. While these are regrettable, they merely reproduce in small the stratification of society writ large.

Task 3

Identify other types of cultural capital in the world around you. What values are they based on, and who gains and who loses from exercising them?

Emergent culture

Discussing contaminated petri dishes at the beginning of the chapter, I said rather emphatically: "In the laboratory, this is a problem. In society, it is a virtue." It is time to address this claim. If we listen to Althusser, it is the hope of the state apparatus to bring all citizens under the sway of a dominant ideology. Everyone would be content with the status quo. The owners of the means of production produce their goods and sell them for a profit and pay their workers enough to live on with a little left over to purchase some of the commodities they made through their labor. As we have seen, this isn't the case because we encounter conflicting ideologies from different ISAs. These tensions between institutions reveal gaps that uncover contradictions and create opposing points of view. These different sets of beliefs and values are variables that in the laboratory cast findings in doubt, making it difficult to reproduce the experiment and get the same results. In society, however, they not only produce conflicts between people with differing views, but they also enliven the world in which we live.

In an open society, as in a dynamic system, perturbations upset the status quo, challenging the stability of the community, if only in minor ways. Currently, Western societies are facing significant disruptions from immigrants moving to escape the violence and political intolerance in their home countries. This is causing problems involving housing, services, security, and employment. In 2016, President Donald Trump promised to build a wall between Mexico and the United States to keep illegal immigrants from entering the country. He has also tried to end immigration from countries he believes are a threat to security. The response to his anti-immigration stand has divided the citizenry between those who believe it is our responsibility to provide humanitarian aid to people in need and those who feel their security and well-being are threatened by outsiders. It has become a conflict between the forces of globalization and nationalism, making it both a scary and energizing time to be alive.

At the same time, there are internal disruptions in many countries over growing inequality in wealth, rights, and opportunities. The conflicts bring disagreements about values and moral stances that have been roiling under the surface for some time. In the UK, there was the vote for Brexit, or Britain's decision to dissolve its relationship with the European Union. In the United States, fights over taxation, increasing debt, women's rights, and fear of terrorism have divided the country in ways not seen since the Civil War of the nineteenth century. At the time of writing, it is unclear how these disputes will be resolved and what the cost will be. But there is little doubt that populations around the world are becoming more active and aware of threats to their personal identity.

Similar events take place in the arts. In 1990, the NEA Four (Tim Miller, Karen Finley, Jolly Hughes, and John Fleck) had funding from the National Endowment for the Arts revoked because of the content of their work (Shockley 2011: 267–84). Their grants were revoked because the performances were perceived to be obscene because of their use of profanity and nudity to challenge religious beliefs, defend gay rights, and condemn sexual abuse. The four performance artists sued, claiming the decision violated their right to free speech. They eventually won the case in front of the Supreme Court, but the NEA ceased funding individual artists. Performance Art, until that time virtually unknown to most of the population, became front-page news and a pulpit for promoting conservative values. More recently, Romeo Castellucci's *Sul concetto di volto nel figlio di Dio (The Concept of the Face: Regarding the Son of God)* was greeted with protests when performed in France:

Outraged by what they consider blasphemy and "Christianophobia", over a thousand protesters (and up to 5,000, according to protest organisers) gathered on Saturday in front of the theatre, hurling eggs and oil at the building and theatre-goers entering it. Two days earlier, police arrested 20 people for breaking into a performance of the play and planting stink bombs. (France 24 2011)

These disruptions occur on the margins of dominant culture, in the spaces that define the avant-garde, experimental, or post-dramatic performances. These forms resist mainstream theatre paradigms and frequently involve a critique of society or issues that affect the lives of the artists. The performances are intended to be provocative by challenging accepted values and beliefs that are perceived to be unjust and oppressive.

Artists explore new forms designed to unsettle the expectations of the audience, encouraging them to question the norms of society. The hope is that by sufficiently challenging the status quo, they will get people to rethink what art is and its relationship to society. Through these experimentations, new subcultures emerge that never fully break with the past, but that influence dominant forms over time. Rap arose from African-American culture as a means of expressing their anger at the prejudices experienced daily at the hands of police and non-black populations. It was gradually appropriated by performers such as Eminem, who gave it legitimacy as a part of mainstream culture. When Lin-Manuel Miranda introduced its rhyming and rhythms in *Hamilton*, critics heralded it as a radically new form of musical. In fact, it was taking a popular form and adapting it to a conventional form. It was innovative but with a significant debt to the past. Nevertheless, what was once viewed as an anti-social form altered cultural expectations of what was acceptable. *Hamilton* is notable not only for the use of rap but also for altering the way history can be

---

Task 4

What new forms of artistic expression (not only theatre and performance) have occurred in your lifetime? Where did they originate? What was the impulse that gave rise to the new cultural forms? Did they become mainstream or did they die out?

---

taught. I know of young people who hated history until they heard or saw the play. Their interest in American history was rejuvenated.

Culture is constantly changing because of resistance to social norms. These disruptions come from progressives and conservatives. While the center attempts to maintain certain values and beliefs, those relegated to the peripheries of society engage in struggles to overthrow what is normative. In this way, cultures evolve, incrementally or radically. New memories are consolidated in minds and societies to be memorialized or denigrated. What is important is that things change. There is always energy roiling beneath the surface, seeking opportunities for expression in new forms.

Culture redux

We have come a long way from where we started, using a petri dish as a metaphor for containing, nurturing, and reproducing a culture to a discussion of memory, plasticity, expertise, and emergent cultures. In a sense, the movement has been from the external to the internal, how societies reproduce themselves in the synapses of memory, and back again, to how artists construct a relationship with social norms. Along the way, we have explored different kinds of memory: explicit and implicit, working and long-term; we've touched on types of collective memories, such as memorials and museums. This chapter concludes with a look at the dynamics of cultural change and the relationship between the center of society and those who live on the peripheries.

The Italian Futurists, part of the European avant-garde that emerged in the first quarter of the twentieth century, sought to obliterate the past within and yet outside of Italian society. Filippo Tommaso Marinetti, publicist for the Futurists, condemned Passéists, those who were complacently living with institutions that were rooted in the past. To embrace the future, they called for an end to religion, museums, educational institutions, and other institutions that were mired in history and tradition. They embraced dynamics of new technologies, speed, and the machine. They embraced war as "the only hygiene." Their desire to destroy the past was doomed to fail because you can't be rid of the past and still live in the present. Memory remains, influencing and directing the choices we make and the future we embrace. From another perspective, they couldn't fail. Societies are dynamic, continually in flux, changing with the pace of life, seeking new ways to respond to discontents and changing circumstances. But this is true for the man on the street, those holding public office, and for artists as well.

Such is the life of culture, always in motion as the need to express creative drives seeks new ways that will regain, reform, or change our relationship with the past. Regardless of motivation, memory remains. We cannot escape the past. Memory doesn't work like that. Rather, within the boundaries of our history, individual, and social, the drive to express our contents or discontents with the world as we see it drives us to push the boundaries of how and what media we use can communicate. Through that struggle innovations in artistic forms emerge. Artists do not create out of thin air but are responding to the culture around them: past, present, and future.

3

The Worlds of Performance

Isaac Newton was sitting under a tree, or so the story goes, when an apple fell either on his head or some distance from him—I'll let you decide. It was from this experience that he began to develop a theory of gravity. This narrative is reduced to its barest essentials, ignoring the dynamics of space that gave rise to his experience. All that is needed to visualize the event is a man sitting underneath an apple tree and the distance between the apple and his head or ground (again, I will let you decide). A lot is excluded from this description. Where is the tree: in an orchard, a field, or his backyard? Was there a reason for the apple to fall? Was it its ripeness? The force of the wind? What was he doing there? Was it a habit? Was he resting after a long walk or doing mathematics? Was it in Lincolnshire, where he was born, or in London, where he died? These details do not affect in any significant way the lore of his discovery, but they are significant to a discussion of space and the events that led to his discovery.

It also says something about the concept of space. The description of the discovery of gravity assumes the ground is a bare surface, the limb from which the apple fell a support, and the distance between the apple and the ground of no real consequence. It is a sketch without reference to landscape or context. "A surface that can only be arrived at through a process of abstraction and reconstruction: by excising every variation or particular from the environment of which it is a part, remodeling it as a piece of furniture or scenery, and reconstructing the scene by imagining each piece placed on a pre-prepared and absolutely featureless floor" (Ingold 2015: 39). In this chapter, we will move from the abstract to the material as a prelude to discussing the space of performance. In the process, we will need to discuss the nature of affordances, some of the socio-political divisions of space, and the ethical implications of those divisions. Finally, we will come to the imagined and possible worlds of the theatre in the reconstruction of space and how those decisions affect the experience of the actor and spectator.

Geo-political divisions of space

From outer space, the earth remains a blue sphere, interrupted by land masses, rotating on its orbit in the vastness of the universe. Down on the planet, this image is quickly subsumed by the crises of immense proportions that confront us. The partitioning of the world that once seemed reliable has lost its moorings, and borders are becoming porous. Regional conflicts seem intractable, as every step toward resolution is met with another insurgence from those who feel their beliefs, rights, and sovereignty trampled on. With wars come inevitable displacements of population, as those caught in the middle and fearing for their lives seek refuge in foreign lands, forsaking their homes in hopes of finding safety for themselves and their families. The influx of immigrants happens at a time when the fear of terrorism, the disruption of social services, and the impact of a global economy drive some countries to seal off their borders. Fear that traditional ways of life are eroding has encouraged extreme nationalists to reclaim their birthrights, giving rise to increased racism, nationalism, and xenophobia.

Concurrently, the global market and lack of regulation are increasing economic inequality through deprivation of the poor and the amassing of wealth. Business decisions focus on the dividends paid to shareholders rather than on providing the workforce with a living wage or concern for the future health of the planet. While businesses expand to foreign countries and officials purchase massive estates, workers, particularly in developing countries, are asked to work long hours for minimal wages under horrific working conditions. In addition, offshore shelters are created to avoid paying taxes, putting strains on governments and their ability (in some cases their will) to provide social services and maintain their infrastructure. More of the burden for sustaining the commonweal is being placed on those who can least afford it, creating discontent that is frequently displaced on to those with even less, giving rise to racial and ethnic tensions, pitting neighborhood against neighborhood. When riots break out, the racial majority often attacks minorities, burning and looting their stores. Distrust of each other is intensified, creating invisible borders that are crossed at your own risk.

The result is a reorganization of space arising from conflicting desires to be part of a global economy and to conserve traditional beliefs and values. Nationalist parties seek to close the borders to stem the tide of immigration, while corporations seek to outsource production to countries with fewer protections for workers. Business relationships depend on the free movement of capital, exchange of goods, and reciprocity in sharing expertise across national boundaries.

Concurrently, climate change remains a point of contention despite overwhelming scientific evidence. Islands in the Pacific are seeking to relocate their entire populations in the face of rising tides. Industries in developing countries operate without concern for the environmental impact or the effect of pollution on workers. Water is becoming a scarce commodity. At the time of writing, South Africa is facing an end to its water supply. The proliferation of herbicides and pesticides is endangering animals, not only threatening the biodiversity of the world but also destroying insect populations needed for the propagation of grains, fruits, and vegetables. As these problems intensify, there will be greater pressure for populations to relocate to areas of the world where they can survive.

Each of these problems reconfigures how we think about space. The maps that show national boundaries no longer seem adequate to the world in which we are living. It might be better to think of currents, flows of people, flows of capital, or the flows of environmental change.

Considering these global crises, we might ask ourselves what we want to say with our theatre. The worlds we create on stage inevitably reflect the world outside our doors. This isn't to say that we should take on immense crises in our performances. These crises impact each of us in different ways. There is a range of arenas to explore, from the global to the psychological. There are external spaces of political and social engagement, and there are interiors of the mind.

Task 1

In a group or individually, look at the institution you are part of and ask how it is divided spatially. In addition to thinking about where classrooms are in relation to where food is served and students live, think about invisible markers. For instance, at the university where I teach there is a wide diversity of students, both national and international. Because of language problems, the students from Asia tend to self-segregate; and for other reasons, African-American students do as well. Are there similar divisions of space where you are?

An Indian Room by Ariane Mnouchkine, Hélène Cixous, and the Théâtre du Soleil is an exploration of how a woman is to respond to the patriarchal violence of the *Mahabharata*. Sarah Kane wrote a harrowing look at a mind in crisis in *4.48 Psychosis*. These distinct plays address the times we live in from very different perspectives and in different spaces. We will return to these issues later, but first we need to encounter the question of space.

From the bird's eye view to the street view

Tim Etchells, director of the British theatre company Forced Entertainment, discusses their approach to work in a video titled *Making Performance* (Etchells et al. 2008). It bears being quoted in full.

> Sometimes when we're working, we get stuck. And very often when we do that, we end up walking around in the city. Talking and trying to work out what we could do next. And very often in that situation we end up looking around and asking, "Why all the things that we see there in the city aren't in the performance." A drunk running out of a pub; a tree with a cassette tape caught in the branches; or a crossroads where there's three nightclubs, and the music spilling out of them mixes up to make a noise. Or maybe it's just a kid walking slowly on the parapet of a bridge. How could you make a performance that really spoke about the world that you live in now? What kind of things would it need to have in it? What kind of performance would you make if you lived in a city like this one? A city where the motorways and the television signals meet. What kind of performance might you make if you were brought up in a house with the television always on. What kind of performance could speak properly about the times that you are living in? In

all their complexity, simplicity, or whatever. Or you could ask a different question. What would you like to see on a stage or in a performance space? What would you like to see put in that place that you haven't seen there before? What would make you smile or laugh to see it there? What would scare you. Other times when we're working on a piece, and we're unsure what to do next, we spend ages just staring at the set we have built, playing records and thinking what might happen next. (Etchells 2008)

Etchells describes a very different relationship to space. The members of Forced Entertainment are not looking at the divisions from a global perspective, however they are defined. Rather, they are engaging with how people live. This does not mean that they are not aware of larger political and cultural questions. Talking about a 1986 piece, *Let the Water Run Its Course*, Lyn Gardner, critic for *The Guardian*, writes: "We watched a beautiful, bleak show set in a Thatcher's Britain where women wept, men fought and the constant rain suggested a country drowning in its own dereliction. It was driven not by narrative, but by details of 80s life" (Gardner 2009). Their confrontational, often funny performances focus on communicating how cultural and political issues affect the lives of those living in Sheffield, a post-industrial city in the north of the UK where they create their work. At the same time, they are confronting the conventions of theatre. In Gardner's article, Etchells is quoted as saying, "We don't hate theatre. We're gripped by it—and its liveness. We love its codes and conventions, but we are also frustrated by them and wage war on them." Living within a cultural system that is nested in a political reality, they challenge conventions in order to question the ways in which we live.

Looking at the ways in which space is defined using a geo-political perspective can be described as a god's eye view, or a bird's eye view. That is, it is as if we were looking down from above and describing what we see. Forced Entertainment rejects that paradigm to engage using a street-level or a phenomenological point of view. **Phenomenology** is "the study of structures of consciousness as experienced from the first-person point of view" (Woodruff Smith 2013). When Etchells and company work on a show, they don't turn to the newspapers but walk the streets, thinking about the way people live and how that can be reflected in their performance. For Tim Ingold, this is what distinguishes **between** from **in-between**.

The metaphor Ingold uses is a river defined by the flow of the water *between* its banks. Understanding the river by its banks is dividing the world from a bird's eye view. "'Between' articulates a divided world that is already carved at the joints. It is a bridge, a hinge, a connection, an attraction of opposites, a link in a chain, a doubled-headed arrow that points at once to this and that" (Ingold 2015: 146). It is the result of analysis and is one way we learn. There is another way: engaging the world as we experience it. "'In-between', by contrast, is a movement of generation and dissolution in a world of becoming where things are not yet given … Between has two terminals, in-between has none." Ingold is interested in becoming rather than being. Being marks a conclusion reached, defined as the banks define the river. Becoming is the flow of the river, in midstream where the current is the fastest. Ingold cites philosopher Michael Polyani: "Tearing away the paper screen of graphs, equations and computations, I have tried to lay bare the inarticulate manifestations of intelligence by which we know things in a purely personal manner" (Ingold 2015: 148). He is not denying the value of "book learning," but understands that the process of learning algebra also takes place in midstream, being an act of generation. The focus is not on a philosophy of abstractions, but on the phenomenological experience of "a world of becoming where things are not yet given." Who we are is not who we become, but who we are becoming.

Space is not a static arrangement of boundaries and objects, but a dynamic environment that comes to mean as we experience it. When Forced Entertainment goes to war with the codes and conventions of theatre, they are resisting what it has become in an effort to see what it can become and, in turn, what they are becoming as artists and members of society. As they combat the traditions of theatre, they are also combatting the codes and conventions of society in an effort to see how they are trying to define us, to make us into beings rather than encouraging becoming. They walk out onto the streets of Sheffield because theatre is not something that happens on a stage in front of an audience, but thrives as an ongoing conversation between the performer, the spectator, and the socio-political environment in which we live. It is something that they know not through analysis, but through seeking ways to ensure communication through confrontation, humor, and passion about the "complexity, simplicity, or whatever" of the world in which we live.

Attention and affordances

Affordances were briefly introduced in the first chapter. Here we will look at them in somewhat greater depth with a focus on relations to space and add another cognitive function: **attention**. We attend to the environment around us from birth. Maxine Sheets-Johnstone, a philosopher, dancer, and evolutionary biologist, asserts that from the moment we are born, we are moving and learning from our experiences in the world. She identifies three abilities that seem to be innate: joint attention, imitation, and turn taking (Sheets-Johnstone 2000: 343). The child, by focusing on the caregiver, perceives that person's movements and attempts to reproduce them. After attempting to copy an action, attention is returned to observation, creating a feedback loop, or trial and re-trial as long as attention between them is maintained. Through these reiterations, we learn the neuro-muscular coordination that allows us to learn the skills necessary to cope with an increasingly independent existence.

From time immemorial, parents and teachers have been telling their children and students to pay attention. It is not that the young people are not attending to something; they are attending to the wrong thing. What they are being asked to do is to refocus on the subject matter at hand. Scientists define this phenomenon into three aspects of attention: **alertness, orientation**, and **executive** (Peterson and Posner 2012: 3–4). Alertness refers to signals in the environment that draw attention to themselves, while orientation refers to how we position ourselves in relation to what is being perceived. The executive involves selecting one thing to focus on and ignoring other stimuli in the field. In an experiment involving selective attention, six participants pass a basketball. The subject is asked to count how many times those in white shirts pass the ball. The number is fifteen, but what is surprising is that a majority of subjects are not aware that a woman in a gorilla suit had walked through the middle of game (Chabris and Simons 2010: 5–6). The directions of the experiment alerted the subjects to what they should attend to, they oriented their attention to the act of counting, and the act of counting privileged those actions while suppressing the sensory stimuli that would have made the gorilla visible.

This activity takes place in two areas of the brain. Maurizio Corbett and Gordon L. Shulman have identified two systems that are involved in exercising attention:

We propose that visual attention is controlled by two partially segregated neural systems. One system, which is centred on the dorsal posterior parietal and frontal cortex, is involved in the cognitive selection of sensory information and responses. The second system, which is largely lateralized to the right hemi-sphere and is centred on the temporoparietal and ventral frontal cortex, is recruited during the detection of behaviourally relevant sensory events, particularly when they are salient and unattended. (2002: 201–2)

The first system is associated with top-down processing and involves memories of past experiences that establish expectations about what will be most important to focus on (see Figure 5). The latter is bottom-up and focuses on selecting and processing stimuli from the environment. As Corbett and Shulman indicate, these are not autonomous systems but are in conversation with each other, each influencing the other. It is impossible to privilege one over the other because they work in concert with each other, one taking precedence over the other only when circumstances dictate.

How we perceive is determined to a certain extent by expectations based on our intentions and memories of previous experiences. These expectations determine, to a certain extent, our intentions, or what we are planning to do. Certain perceptual information is selected because it is considered suitable to the intended action. Andy Clark, in *Surfing Uncertainty*, talks from an AI perspective and focuses on neural predictions as the primary force in defining the focus of attention, putting less emphasis on what is perceived and more on "the ceaseless

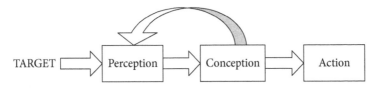

Figure 5 *Attention: Relationship between top-down and bottom-up processing (Bodenhausen and Hugenberg 2015: 2).*

anticipatory buzz of downwards-flowing neural prediction that drives perception and action in a circular causal flow. Incoming sensory information is just one further factor perturbing those restless pro-active seas" (Clark 2016: 52). Placing emphasis on intentional actions, he argues that we anticipate what is going to take place and base what we are going to do on those predictions. He does recognize that input from the world can disrupt these predictions, calling them prediction errors. They are the perturbations that roil "those restless pro-active seas."

It feels right that the brain determines our interactions with the world; this does make sense when we talk about habitual actions, such as brushing our teeth. There are situations, however, such as driving a car, when our attention to incoming sensory information needs to be at least equal to our predictions about what is going to happen. We need to be alert when a teen riding a bike suddenly swerves in front of the car. The same is true of going to the theatre. A major tenet of this book is that the aesthetic experience is based primarily on *exceeding* expectations. It would be a dull experience indeed if the performance we were watching matched exactly our predictions of what would happen.

This relationship between what we expect and perceptual information is what it means to be embedded in the world. Previous experiences allow us to anticipate what will take place, but they are balanced by the information coming from our environment. It is a passive–active relationship. We act, evaluate, and act, but not necessarily in that order. It all happens so quickly and most frequently unconsciously so experience seems to flow. That is, except for those moments when what is happening does not match what we thought would happen. Then we need to assess the situation and devise alternative forms of action. What we are seeking to understand and evaluate is what the world affords.

Task 2

Think of a particularly memorable experience in the theatre or any other art, or if none come to mind, another significant moment in your life. What happened to make it stand out? How was it different from what you expected?

As we saw in Chapter 1, affordances define the relationship we have with the world: our perceptions in conjunction with our intentions allow us to act with confidence or at least a sense that we will succeed. Task 6 in that chapter asked you to trace your movements getting to your classroom, including elements in the environment and the abilities that allowed you to arrive at your destination. Here we will expand and complicate that idea.

Anthony Chemero, who advances Gibson's theory, argues that the idea of affordances does not refer to properties of the environment or the animal. Instead he sees them as relations between people and the world in which they live; he names them *dispositions*. "Affordances are neither properties of the animal alone nor properties of the environment alone. Instead, they are relations between the abilities of an animal and some feature of a situation. They are not easily localizable physically but are nonetheless perfectly real and perfectly perceivable" (Chemero 2003: 191). If affordances are perceived as properties of the world, the expectation is that they are more or less permanent features. Chemero argues that there are aspects of our engagement with the environment that vary, such as changes in the weather, that nonetheless dispose us to act in certain ways.

When going to the theatre, we may expect that there will be seats and that they will be faced toward the stage where the action will take place. These are affordances in Gibson's sense of the word. These elements were present when I went to see *Les Atrides* by the Théâtre du Soleil, but there was an additional element. On walking into the theatre, I became aware of a sound filling the space that can only be described as a drone. It was a continual presence in the space that set a tone for the tragedies to be performed. This was not a property of the space but a quality of the environment that affected my relationship with the space. This is what Chemero means when he uses the word affordances: "Affordances, I argue are relations between particular aspects of animals and particular aspects of situations" (Chemero 2003: 184). Or in his preferred language, this is what he means by dispositions. The Italian Futurists, although this may be apocryphal, sold more than one ticket for a seat and put glue on the seats to provoke the audience and jolt them out of their "passéist" ways.

Tim Ingold takes Chemero a step further in thinking about affordances. For him, the idea of a relationship that defines a disposition is too static a conception of the relation between those who act and the world they engage. Citing philosopher Jan Masschelein, he writes:

> To walk, as Masschelein puts it, is to be commanded by what is not yet given but *on the way* to being given. It is not, then, that the walker's attention is being educated; rather the reverse: his education is rendered attentive, opened up in readiness for the "not yet" of what is to come. (Ingold 2015: 136)

Paying attention utilizes our past experiences to move with confidence based on predictions about what-is-to-come as we explore the environment. We are ready for the experience of living, just as an actor stepping from the wings onto the stage is prepared for that evening's performance. We act on a set of changing predictions in response to the events of the day from the moment we get out of bed to getting ready to fall asleep at night.

Engaging space, whether by an actor, stage hand, technician, or audience, is an embodied action based on previous experiences that anticipate moving into the world in all its "complexity, simplicity, or whatever." Space is not an abstract concept in this context but the realm of experience in which we are embedded, as embodied beings, and into which we extend ourselves, enacting our intentions. It is from this perspective that we move into the question of space in the theatre.

Possible worlds

Theatre always has a problematic relationship with reality. No matter how realistic productions become, they are always understood to be outside of everyday life. This is both a blessing and a curse. Jonas Barish, in *The Anti-Theatrical Prejudice* (1985), traces the lineage of prejudices against the art form. Primary motivations for the distrust of the theatrical are the actor's ability to shift from character to character, making the true identity of the actor amorphous; more to our purpose is its dependence on illusion, not only by the actor but also in the whole of the *mise en scène*. Interestingly, the things that give rise to a distrust of the theatre are the very things that provide it with power. The ability to assume different personas and to create spaces that are not identical with the everyday allows theatre to comment on the world through metaphor and analogy. Whether it is the investigation of psychological relationships, a commentary on political realities, or a disruption in the fabric of the theatre itself, the art form creates a dialectic between the images on stage and the experiences of the spectator.

The goal is to create a **possible world** for the production, "a complex network of relations between the factual and the non-factual, the actual and the virtual" (Ryan 2012: 18). Possible worlds use the organization of space to create an environment in which the reality of the performance can unfold. Josef Svoboda, the Czech designer and director, popularized the concept of the *scenographer*, a designer who designs all aspects of the performance, and was instrumental in bringing new technologies to the stage through his work with Laterna Magica. "When I sit alone in a theatre and gaze into the dark space of its empty stage, I'm frequently seized by fear that this time I won't manage to penetrate it, and I always hope that this fear will never desert me. Without an unending search for the key to the secret of creativity, there is no creation. It's necessary always to begin again. And that is beautiful" (Svoboda 2018). When Svoboda begins his process, he visualizes the stage space as darkness that needs to be penetrated by creativity in the act of designing.

This is similar but different to Peter Brook's concept of an empty space. For Brook, the actor crossing the stage is an act of theatre. Svoboda begins to think about a space before the arrival of the actor, even before the introduction of light. Staring into the darkness, he begins to imagine how the space can be organized to communicate the vision of the text that he has developed with the director. His goal is to create a correspondence between the play, the design, and the work of the performers. This process includes deciding what kinds of technology are needed to realize the vision. This may mean parts of the stage move during the production or that film and video are incorporated into the design, "but sometimes, if it seemed right for the production, his sole decoration could be as simple as the silhouette of skyline" (Martin 2002: 1). For a production of Chekhov's *Three Sisters*, he "strung groups of silvery cords between stage and lighting grid, which were instantly transformed by light from inside to outside, from solid wall to shimmering space" (Jays 2002: 1). The interplay of the design elements transformed the stage image to correspond to the unfolding of the text, underscoring through other means the content of the play.

There are as many possible worlds as there are productions. There are traditional sets, such as described by the stage directions for Joshua Logan's *Red* at the Arena Stage, which call for "a fully functioning studio, complete with paints, skylights, canvases, and industrial lamps" (Gilbert 2012: 1). The objective is to design a set that is as close to the world of everyday life as possible, down to running water and functioning hot plates. Less demanding of realism, sets for classic plays free the designer to create spaces that speak of a world separate from everyday reality. Peter Brook, on directing *Hamlet*, said, "I felt that a new and concentrated version of 'Hamlet' could freshen and bring to the surface things that have gotten partly buried through our great familiarity with the play … Underneath, there is very often a purer and stronger and deeper work that today is more relevant. And you can reveal it by delicately removing the superstructure" (Riding 2000). Chloe Obolensky responded to Brook's concept with an atypical set. Gone are the battlements and halls of Elsinore. In its place are "an orange-red carpet, two stools, two skulls, some cushions and the peeling rear wall of the theater" (Riding 2000). (The theatre referenced may be the BAM Harvey, part of the Brooklyn Academy of Music.) There are a multitude of different approaches to design in between the naturalist and the minimalist, but all attempt to create a world, providing the audience access to the play defined by the work of the creative team. At the same time, the scenography must allow the director the freedom needed to realize the interpretation. It must also support the work of the actors by creating an environment that is conducive to physicalizing the actions they need to perform and the relationships they need to develop.

Not all possible worlds are created on a stage. Punchdrunk purchased a warehouse in the Chelsea District of New York City for their production of *Sleep No More*, an immersive theatre piece based loosely on Shakespeare's *Macbeth*, Hitchcock's *Vertigo*, and Daphne du Maurier's *Rebecca* (Brantley 2011: C1). Each of the five floors of the building was dedicated to a different environment. One depicted a turn of the century town in great detail, while others were an insane asylum, a hotel lobby, and a forest made of dead branches. The audience entered through a darkened hallway and followed a maze up the stairs to a bar where drinks could be bought and music played. Given a mask and taken to an elevator, spectators were dropped off on different floors and left to explore one environment before taking the stairs to another level.

Conni's Avant-garde Restaurant "staged" a production of Shakespeare's *Twelfth Night* in the Meatpacking District of New York City. Audience members were told to assemble at different locations in the district where they were met by a guide who led them through the streets of the city.

> Our tour guide for the evening, Frankie, played by Rachel Murdy—who also conceived the piece—didn't just recreate Feste (the court jester of *Twelfth Night*) as an innocuous New Yorker. Oh, no. She transformed the character into an institution of the neighborhood. Referring both disdainfully and endearingly to the constantly evolving district as "MePa," [providing a] history of the Meatpacking District from the arrival of Henry Hudson to the gay club scene of the 70's. (Rainville-Tomson 2012: 1)

Different groups had different leaders and met at different locations. As the audience was led through the area, different scenes from Shakespeare would suddenly emerge from the darkness. Toby would appear on the street and argue with Maria, who was talking from the second-floor window of a neighboring building. A veiled Olivia lived in a room of an upscale hotel, grieving in the dark as the visitors/spectators drank champagne by candle light. At the end of the evening, the various groups, who had crossed paths throughout the performance, met at the entrance to the High Line (a pedestrian walkway on the remains of an elevated train line) for the last verses of the play and a slice of cake.

To offer another example, Improv Everywhere generated a flash mob of 207 participants. They went into Grand Central Station Terminal in 2008, moving throughout the main lobby. When the clock struck 2:30, they froze for five minutes:

> It was fun to see all the different choices people made for their frozen moment. I didn't give any instructions in advance. I just told everyone to be doing something realistic and not jokey. One guy dropped an entire briefcase full of papers the second before he froze, leaving his papers scattered before him for five minutes. Many froze midway through eating or drinking. A few froze while taking off a jacket. One couple froze kissing. (Improv Everywhere 2008: 1)

At the end of the time, they unfroze and left the building. A far reach from the naturalism of the design for *Red*, a space where people rush through on their way to work or home was transformed into a theatrical world where people stay perfectly still, and then resume their day as if nothing had happened.

Possible worlds are sites of imagination, whether a proscenium stage or a found space. In the five minutes the participants were frozen in Grand Central Station, it ceased to be the train terminus for commuters taking the Metro North trains into the City or back to the suburbs and became an unfamiliar place where people froze for a period of time and then returned to their everyday lives. Similarly, an orange-red rug and the peeling back wall of the theatre became a castle in Elsinore. But possible worlds are only convincing if they become spaces in which the unexpected (regardless of how realistic)

Task 3

Think of a significant production in your experience. Define the possible world that was created for that event. It need not be traditional theatre. A more difficult choice, perhaps, would be to think of a concert or a dance performance.

---

can take place without undercutting the significance of the event or destroying the illusion being created.

Space and performance

The classifications of performance spaces are quite small, but the variations within each category are virtually infinite. Outside of the triumvirate of traditional spaces—proscenium, thrust, and arena—there are site specific spaces that can be indoors or outdoors, and public or private. Theatre can take place in museums, warehouses, churches, town squares, open fields, abandoned factories, swimming pools, or in one instance, a closet. The one thing that remains, although the boundaries are being challenged, is the distance between the performer and the spectator. Even in the most participatory theatre, the actors have a knowledge and the ability to control what happens, however attenuated it might be. In this section, the focus is on how the production team engages the space, beginning (if there is a beginning) with the entry into the space of rehearsal and through the different phases of rehearsals that lead up to opening night.

Engaging with space in the rehearsal process is full of discovery and anticipation, especially, as is often the case, if the preparations do not take place where the work will be performed. Practitioners entering an unfamiliar space will frequently look about them and move through the room, sensing its dynamics, wanting to get comfortable as soon as possible in order to concentrate on the work. They check out the dimensions of the room, side to side and top to bottom; the acoustics, whether there is resonating echo or if the sound is dampened; the lighting, neon or incandescent; and the color of the walls, white or black. The stage manager and director arrange the furniture, creating an approximation of the set and deciding where to place the table for text work. There might by a taped outline of the ground plan and real or substitute set pieces. These explorations might happen consciously or unconsciously as the performer begins to warm up or greets other members of the production team. Regardless, it is a process of attending to the space, of becoming aware of its potentials and limitations, of its affordances and dispositions, because this is where the work is going to take place until they move into the theatre.

As the rehearsal process progresses, the performers attend more to the tasks required by rehearsal and the work done in their off-hours. The qualities of the space will recede to the background unless there are changes or the limitations of the room become obstacles to the work. New spatial concerns will arise as the distances between the performers and the location of set pieces become more prominent. An explicit aspect of the process is defining movement patterns and images, and their relationship to communicating the creative vision. The director and actors will need to decide when to bring the characters together and when to separate them, when they should sit, stand, kneel, lie down, or climb on the furniture. These issues are significant because they can help the performers define their relationship with each other and clarify the action and the intensity at which it needs to be played.

Props and costume pieces will later be introduced necessitating new explorations to discover how they can be used to further the action, redefining the relationship to space, and determining what they may afford the actor in revealing character, relationships, and the intensity of the action. They may be used to define the psychological dynamics of the production or promote the intentions developed by the production team. Brecht explored props to see how they could be used to define a *gestus* that underscored the political commentary of the text. If it is a period piece, do the women need to wear corsets? If so, when do you bring them into the process? They change how the actors carry themselves and their breathing patterns, and they change how actors understand their role in the production. What kind of footwear? Sneakers (trainers in the UK) walk differently than leather soled shoes, let alone boots or high heels. Each of these decisions provides the performer with information about their character, the style of the production, and the image they present to the audience. When I directed Steven Sondheim and Hugh Wheeler's *A Little Night Music*, a member of the chorus rebelled against wearing a corset. On opening night she announced that hers was "lost." As she prepared for the performance, she noted that the costume still fit. Another member of the cast quipped, "The only difference is that instead of an hourglass, you look like a stopwatch."

As the performers become accustomed to the set, props, and costumes, they will begin to explore their options in relation to the action, character, and intensity of the scene. The flash mob of Improv Everywhere only required that the participants turn up at Grand Central Station in whatever clothes they decided to wear that day. As noted, one person chose to bring a brief case that opened just before the start of the freeze, scattering papers across the floor. He froze as he bent down to pick them up. This simple choice added to the event, providing a different level from those who remained upright, adding a dynamic that would otherwise have been missing. Some experiments will be discarded almost as soon as they are tried, and others will be elaborated or transformed becoming defining qualities of the character and action. In *The Grapes of Wrath*, Granpa Joad has a limp. When I first explored this aspect of the character, the movement was seen to be painful, creating a certain way of crossing the stage. The director felt this choice was not in keeping with the production. He suggested I think of the limp as something that happened a long time ago, and that Granpa had learned to compensate for it. This freed me to focus on the action, developing relations with the other characters, and a better understanding of the patriarch's personality.

Each of these choices affects the spectator's experience and is interwoven into the production. The event might be as simple as "go to Grand Central Station and, at 2:30 p.m., freeze for five minutes" or as complex as Robert Wilson, Phillip Glass, and Lucinda Child's *Einstein on the Beach* (Daily Motion 2012), a four-and-a-half-hour performance without intermission involving music, dance, text, and spectacle. On the one hand, the participants in the flash mob were free to determine their own performance. On the other, Wilson's direction and Glass's compositions require precise timing and are rehearsed down to the last detail. But in each instance, they are exploring the use of space in performance.

Task 4

This description of the process includes aspects that resonate with Gibson's concept of affordances and Chemero's idea of dispositions. Identify those that can be attributed to Gibson and those more appropriately seen as reflecting Chemero's theory. Think of a performance you have seen or participated in. Can you single out aspects of that production that relate to affordances and those that are dispositional?

Toward the end of the rehearsal process, the company will move to where the performance will take place. It is to be expected that the production will suffer. New elements are introduced, such as the actual costumes, props, and set pieces, as well as the lighting and sound designs. Equally important is becoming acclimated to the new space. All the things that were encountered on the first day of rehearsal will need to be explored once again: the volume of the space, the acoustics, and the relationship of the set pieces to the overall architecture of the design. These elements affect the relationships that have been developed among the performers and the production team requiring transformations, however minimal, to the production that had been shaped during rehearsals. The atmosphere has changed from one of excitement at beginning a new project to the introduction of new energies with different foci. Lights are being cued, costumes adjusted, prop tables organized, and the stage manager organizes communications between the designers and technical staff.

What has just been described is a more or less traditional theatre production. Other theatre performances, as we have seen, require much less preparation and have a different relationship to space. Ben Vautier, a Fluxus performer, enacted *Solo for Violin* in New York City, sitting in a chair tuning his violin. He is bound by strings that connect him to the wall of a building and there is a sign that says: "Fluxus/Street Theater/ Free." Next to him was a suitcase with bags of undetermined contents.[1] A very different kind of preparation is needed to create this performance. Nevertheless, it requires an exploration of space, making use of affordances supplied by the building and the street, and the dispositions created by the weather, passers-by, and the noise of the traffic. All of these determine the performance, in a sense making it possible, and create the effect space plays in the reality of *this* possible world.

---

[1] An image of the performance is available at https://www.moma.org/collection/works/127431.

There is another aspect of space that we have not yet covered. That is, the space between the performers and what that affords them. This will be discussed in the next chapter. It is worthwhile mentioning at this time that the relationship between characters is a material and metaphysical distance that is nested in the possible world. Like a Matryoska doll, the work of the performers is nested in the space of the design, which is nested in the volume of the theatre or wherever the performance is presented. Where this analogy falls down is that the nested dolls are separate entities, while in the theatre this relationship is part of a **meshwork** in which each element is inextricably linked with the others, creating direct and interwoven connections between the possible world and the cognitive processes of the performer and spectator.

Space and the spectator

Like the performers, spectators have a physical and metaphysical relationship with the performance. It is typical to think of the audience as passive bystanders, as Tom Stoppard reminds us in *Rosencrantz and Guildenstern Are Dead*.

> (**Ros** *leaps up and bellows at the audience.*)
> **Ros**   Fire!
> (**Guil** *jumps up.*)
> **Guil**   Where?
> **Ros**   It's all right—I'm demonstrating the misuse of free
>   speech. To prove that it exists. (*He regards the audience,
>   that is the direction, with contempt—and other directions,
>   and front again.*) Not a move. They should have burned
>   to death in their shoes. (Stoppard 1967: 60)

The spectators didn't move for Stoppard's character because they knew the alarm was not a real one. They knew it was scripted, just as they knew he was not "demonstrating the misuse of free speech." That too was part of the play. The audience comes to the theatre with a set of expectations. They anticipate that they will have a seat and that they will be facing the stage. They know they will be safe, because if there were strong noises, fog machines, or strobe lights, there would be a sign warning them in the lobby. This wasn't always the case.

In the medieval period, the audience would gather in the Nave, or open area of the cathedral (there were no pews). Between the pillars that supported the roof, there were *mansions* from which actors would appear, presenting different stories from the Bible. The audience would move about from one to the other moving back so the actors had room to perform. In Shakespeare's time, part of the audience would stand in the pit, while the wealthier clients sat in seats. The players would address the audience directly, sometimes taking them into their confidence through asides. In the eighteenth century, performances moved into theatres, lit by candles. To be able to see the play, they were kept lit throughout the performance. The actors had to compete with what was going on in the audience where orange girls were selling food and prostitutes arranging assignations. It wasn't until the advent of gaslight and then electricity that the spectators were allowed to sit in darkness to watch the play.

Each of these conventions required a different kind of spectating, establishing expectations of what the theatre experience would be. Each afforded the audience a different kind of experience and placed different demands on where the focus of their attention would be. In court performances, the attention of the spectators would be as much on who was there and how the king or prince was responding as to what was happening on stage. But there is one constant in each of these different configurations: the performer is separated from the audience by a distance that is respected.

Early in the twentieth century, this separation began to be challenged. Early avant-garde performers began to challenge the audience to get them out of their complacency. The Futurists would harangue the spectators, trying to get them riled up so they would fight back. As noted earlier the Futurists sold multiple tickets for the same seats, leaving it to the ticketholders to fight it out. There are reports that they put adhesive on the seats, gluing the viewers in place. If the event was sufficiently antagonistic, and they often were, fruit was pelted at the performers and all would charge out of the theatre, onto the streets, and into the arms of the police. These strategies were taken up by The Living Theatre, which often performed in venues other than theatres. They invited the audience to come onstage and participate in the ritualized performances, again going out into the streets to protest the capitalist system and government oppressions. Richard Schechner founded the Performing Garage and staged a performance of *Dionysus 69*. Part of the performance included the actor playing Dionysus asking one of the spectators if she wanted to sleep with him. One night, so the story goes, the woman said "yes." They went off and the rest of the performance was cancelled.

Despite these efforts to disrupt the relationship between the spectator and the performer—to penetrate the distance between the two—the tradition of separating the audience from the stage persists. Most theatre productions, even those that strive to undermine the dominance of plays that present a narrative and psychologically based characters, use conventional seating arrangements and divide the actors from the spectators. The audience sits in relative darkness while the performers remain within the confines of the stage. The value of this arrangement is that it helps the spectators attend to what is happening on the stage. Through creating this focus, the hope is that the impact of the performance, whether tears or laughter, will be an optimal, intense experience.

In the second half of the twentieth century, artists dissatisfied with the limitations of the gallery and museum broke with convention and made art that was not lasting but existed in the impermanence of performance. In France, the Situationists, whose chief spokesman was Guy Debord, curated performances in public spaces designed to disrupt the numbing effects of the "spectacle" and encouraged people to engage the space in unexpected ways. In New York, George Maciunas wrote the Fluxus Manifesto in 1963, a loosely organized group of international artists and performers who took theatre into the streets, taking passersby by surprise and creating an audience out of those who stopped to see what was happening.

There are other explorations of the relationship between the performer and spectator. I would like to discuss three before concluding this chapter. One is returning to Punchdrunk's *Sleep No More*, which was introduced earlier; the second is *The Encounter*, a production of the London company Complicité and performed by Simon McBurney; and the third is Katie Mitchell's production of *Waves*, based on the book by Virginia Woolf. The first is a form of **immersive theatre** that disrupts the traditional relationship between performer and spectator by allowing the audience to literally enter into the possible world of the production. The other two maintain the normative separation between the stage and house but require the audience to actively engage with the piece in unusual ways.

On arriving at the Chelsea warehouse, which has been dubbed the McKittrick Hotel, spectators line up outside, like partiers waiting to go clubbing. When the doors are open, you go down a dark corridor. At first, you come to a hatcheck area—purses and backpacks are not allowed in the performance. You then proceed to the hotel check-in desk where your reservation is noted and you are given a random playing card. You are then directed down a dark corridor that turns out be a labyrinthine passageway to the second floor. There you are greeted by a maitre d' who informs you that the performance will begin shortly, and that there is a bar where you can get drinks in a plush 1920s nightclub. Once the audience is assembled, the master of ceremonies begins to call out playing cards. It turns out that you are one of about twenty who have the same one. You go into a side room, receive a white mask that you are told to wear at all times, and you are told that you are not to talk for the duration of the performance. You then cram into an elevator and are taken to one of the five floors. The doors open into an environment that you are at your leisure to investigate.

As you explore the space, you begin to realize that what is in the environment is slightly surreal. On a desk, you find a ledger on which a bird is pinned and with lines drawn to different parts of its anatomy. You follow a path through a stand of tree limbs, encountering stuffed animals before you come to a nurse's station. You find yourself in a medical ward with no patients, but there are medical histories you can read. The experience is eerie because of the music playing and the absence of anyone other than your fellow audience members. Suddenly actors appear in period clothing. They interact briefly and then move quickly away. You follow them but by the time you reach the corner they have disappeared. As more of these events take place, you attempt to create a narrative that makes sense of the chance encounters. But there is never enough information. It begins to feel like you are in a dream.

On occasion, an actor picks you out of the crowd and pulls you into a small room. The first time this happened to me (it doesn't happen to everyone), she told me a story of a young boy who meets a witch in the forest and is told to find a ring if he wishes to find his way out.

The boy looks but is unsuccessful and dies. A whisper in my ear tells me I know where the ring is. I am pushed out a different door into a completely different environment. At the end of the evening, we gather in one room to see the hanging of Macbeth, who swings over our heads. The audience is invited to the bar where you can meet the cast, have drinks, and leave when you are ready.

The experience is surreal, in the sense of blending dream worlds with the reality of your presence. What is relevant to this discussion is that the mask differentiates you from the unmasked performers. While you are free to roam where you will and explore the environments, you are always aware of the performer separated from spectator. The separation is physical (as in the medieval theatre you make room for the performers) and psychic. You never feel that you are in control of what is happening, and the environment is governing your participation. The world of the McKittrick Hotel affords you immense opportunities to explore the worlds they have created. There may be events that are missed as you study the intricacies of the space. You can attend to what is most interesting to you for as long as you like. But there are no answers, no glue that binds it all together, and you are always limited by the contours of the space. As soon as expectations are established, they are contradicted.

A very different approach to the audience is explored in the one-man show *The Encounter*. The proscenium stage is designed to reflect a sound studio, with a table stage right littered with bottles of water, papers, a lamp, and microphones. Center stage, there is a mannequin wearing a set of headphones. As the house lights go down, the audience is asked to put on the headphones that are attached to their seats. The performance is based on photographer Loren McIntyre's trip down the Amazon (based on Petru Popescu's nonfiction account of the journey in *Amazon Beaming*) and encounter with the Mayoruna tribe and "incidentally, his discovery of the source of the Amazon" (Soloski 2016: 1). It is not a straightforward narrative, however. There are layers to the performance. In addition to telling/performing McIntrye's journey, McBurney tells of his own journey in creating the piece and is interrupted by his five-year-old daughter who wants him to read a bedtime story. There are more layers as well:

> And with a host of people he interviewed in preparation for this show, including Mr. Popescu and a variety of neuroscientists, philosophers and environmentalists, who keep interrupting the central story with fragments of theories you've heard them expound earlier in the show. (Brantley 2016: 1)

The audience also sees McBurney making necessary sound effects: insect noises made by rubbing a water bottle and sounds of the jungle made by running across the stage shaking yards of audio tape.

You see it, but what you hear is through the headphones. It is very democratic in the sense that everyone hears the same thing, but you are also isolated because it is all happening (or so it seems) in your head. At one point, his daughter opens the door and begins to speak. Everyone in the audience looked over their shoulder because that is where the sound was coming from. It had a strangely disorienting effect, which was in part why the headphones are there. They create the effect of taking "the hallucinogenic frog secretion McIntyre imbibes ... [and] when the lights are flashing and the sound is flooding and McBurney is chanting and dancing and sweating, *The Encounter* is mood-altering and mind-expanding" (Soloski 2016). The experience is powerful because the aural aspects of the production seem to be happening in your head at the same time your attention is focused on what is happening on stage in what is a very physical performance. I experienced a cognitive disjunction between what was on stage and what I was hearing, such as, for instance, when what I took to be McBurney speaking into the microphone turned out to be a pre-recorded tape. I began to question what was real in the performance, which I believe was precisely the intent: to approximate McIntyre's hallucinations.

The third production I want to talk about is *Waves*, an adaptation of Virginia Woolf's novel of the same name (Brantley 2008: C1). I saw the production in 2008 in New York City. The story is about six friends, equally divided between women and men, as they mature from teenagers to adults and their passion for a seventh person, Percival, whom they idolize but who dies after being thrown from a horse in India. As they grow older, he ceases to be the focus of their attentions, replaced by a growing awareness of their own mortality. The production in the Duke Theatre was again proscenium seating. There was no stage, per se; only the floor of the theatre. In front of the first row of seats, a large projection screen was suspended and set back approximately 12 feet from the first row of the audience. In front of it was a long table with eight chairs; a few microphones; and adjustable, gooseneck desk lamps. Down stage on the right and left are four boxes about 3 feet square and 2 inches deep. A standing microphone is placed on one of the squares on each side of the stage, while the rest contain natural materials, such as stones, grass, and dirt. Flanking the sides of the stage are shelves that hold the props used in the performance.

During the performance, different parts of Woolf's novel depicting the lives of the characters were read into a microphone. At the same time, the other members of the cast created the scene visually and aurally. The assembled elements of the scene were projected in real-time video on the screen. For instance, each new section marking the passage of time is identified by time and place, information that is written on a blackboard by one of the actors. Another actor holds one of the desk lamps, illuminating the blackboard, while yet another actor films the act of writing on the board. A fourth actor sits at the table in front of a microphone, creating the sound effect of chalk hitting the board and then rubbing a whisk broom across the hand to reinforce the sound of the writing being erased. The audience simultaneously sees the actors creating the video and the simultaneously generated audiovisual images on the large screen. The remainder of the production is produced in a similar manner, although the images are usually more complex.

It was disconcerting at first, not knowing where to look. The eye was drawn to the screen, but attention was diverted and shifted to the work of the actors putting the scene together. As the production proceeded, it became easier to shift the focus from one to the other while also listening to the text as it was spoken. Success in the perceptual task mirrored the coming together of the lives of the characters as they matured into adulthood. What was missing in resolving the split attention was emotional identification with the characters. You cared about the story and the beauty of the language but were more entranced by the ingenious ways the scenes were put together for the camera. It came as a shock, as the last scenes of the performance were played out, to suddenly feel a wave emotion, one of great sadness at the movement from youthful promise to the realities of sedentary maturity. I suspect there had been emotional responses all along, but the focus on the creation of the scenes kept them in the background. As my ability to reconcile the staging with the filming matured, mirroring the movement of the play, I felt the realization the characters were coming to, just as suddenly the putting together of the piece faded to the background.

Each of these productions provided the audience with very different relationships between themselves and space. The immersive experiences broke down the separation of stage and house, placing us in the set with the performers. McBurney's performance detached the aural from the visual, creating a sense of disorientation that mirrored McIntyre's own disorientation and the hallucinations he experienced in his trip up the Amazon. *The Waves* challenged the spectator to find a way of combining the making of the video with the projection of it while continuing to follow the story. The brilliance of the performances was their success in creating a world that provided a powerful impact, at least on this member of the audience. One provided the surreal experience of being awake in the middle of a dream; another provided the disorientation of a drug induced trip; the last used technical means to evoke an emotional recognition of the loss of hope that comes with aging.

Conclusion

When we enter the theatre, we do not leave the world of the everyday behind us. Nor do we bring only the stresses and pleasures of the day. Our phenomenological awareness is continuous. The task for theatre and performance is to create an experience that will seduce us into forgetting, if only for a short time, what happened before we took our seats, and will transport our attention into the possible world of the performance. This can be accomplished gradually by letting the décor of the theatre distract us from earlier concerns. Or it can abruptly refocus our attention, like the flash mob in Grand Central Station. If successful, we leave the theatre in a different "headspace." Immersed in the afterglow of the experience, we re-emerge into the everyday life of the streets. If it is an exceptional experience, we may, however briefly, perceive that world differently.

Whether we are taking the subway in New York City or hiking the glaciers in the Alps, we are experiencing space. That is, neither the curving space of Einstein nor the flat surfaces of Newton and Kant but *terra firma*, with its terrains and cityscapes, waterways, and skyways.

It is our home, a *sine qua non* of being alive. Even when immersed in the illusions of virtual reality, we may believe we left our terrestrial bodies, but we are still embedded in space. In Ingold's space, this means "a movement of generation and dissolution in a world of becoming." The element that we have not discussed in our exploration of space is the invisible arrow of time, to which we turn our attention in the next chapter.

4

Temporality

For Albert Einstein, time is a fourth axis distinct from the three dimensions of space. It is not a separate entity but woven into the space-time continuum. Having no dimensionality, there is no beginning or end; therefore, time cannot be divided into past, present, or future. Quantum theory has called this into question. You can determine the location of a particle or the speed at which it moves, but you cannot do both simultaneously. They are not intrinsically linked. These abstract concepts, interesting as they are, are not very useful in everyday life because the vantage point from which they are observed is that of the universe. We need a more down-to-earth concept of time, one that allows us to schedule our lives. To achieve this end, time is divided into years, days, hours, minutes, and seconds; because of this we can make it to school, catch a plane, and be at the doctor's office on time. This change in the way of thinking about time shifts us from a god's-eye view to a bird's-eye view. It is a better way to live our lives, but still does not get us to the everyday experience of time. We need a street view, an eye-to-eye theory. We need to turn to phenomenology.

Edmund Husserl was one of the first philosophers to address the question of time. He was less interested in incremental time than in trying to understand the correspondence between the past, present, and future in our daily lives. His formulation involves three terms: the primal impression, retention, and protention (see Figure 6). ***Primal***

impression refers to our consciousness and perceptual experiences as they vary from moment to moment. ***Retention*** is what remains of the impression, or what we are able to hold in short-term memory. Based on the primal impression and what is retained, we are able to anticipate the immediate future. Husserl called this projection into the future ***protention***. The flow of experience that gives continuity to our everyday lives depends on retaining the immediate past for a period of time but it fades as new primal impressions accrue. Similarly, as time progresses we update our expectations, and earlier anticipations lose their force. In Figure 6, the vertical lines are the primal impressions and the heavy horizontal line is the flow of time. The diagonal lines labeled R refer to the retentions that arise out of the

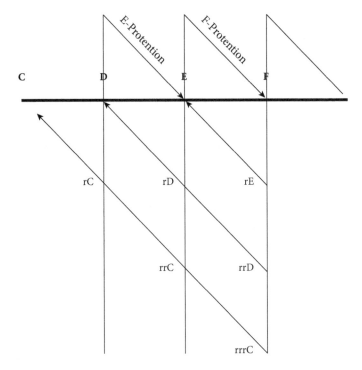

Figure 6 *Husserl's temporal structure of consciousness (Gallagher and Zahavi 2012).*

immediate moment and steadily dissipate as time passes. Similarly, each diagonal labeled P refers to a protention that also loses force as new information arrives from the material world.

There is great value in this theory. Marc Whitmann in his book *Felt Time: The Psychology of How We Experience Time* cites studies concluding that a momentary experience has a duration of about three seconds before it fades from consciousness (Whitmann 2017: 39–58). It can last longer but only if it is held as part of a bit of information in short-term memory. It also explains why time appears to flow in an unbroken stream, creating the continuity of experience in ways that incremental concepts cannot. However, it is less useful in defining breaks in time. Clock time is valuable in measuring time; phenomenological time is helpful in understanding the passage of time. What is needed is a theory that can do both.

Gilles Deleuze, a French philosopher, provides a different approach to conceptualizing the temporal. Instead of the familiar triad of past, present, and future, Deleuze defines four aspects of time: the pure past, particularities, the living present, and generalities. The first three map onto Cowan's diagram of short-term memory (Chapter 2). The **pure past** is similar to long-term memory. It must be retrieved to become conscious, and only when brought back does it have a recognizable form. The past, when it is retrieved, participates in working memory as active data that can be brought into consciousness. Deleuze calls this information retrieved from our memory banks and that which has just passed (Husserl's retentions) **particularities**. The **living present** is associated with the circle of attention, which contains the data needed to negotiate what is happening now. The final term without equivalent in Cowan's description is **generalities.** These are Husserl's protentions, or anticipations about what the future holds, but as the word implies, they lack specificity because we don't know precisely what is going to take place, and over time things can change in unexpected ways.

Thus far, while there are refinements, Deleuze's taxonomy is similar to the other theories that distinguish between past, present, and future. The concept that Deleuze brings to the table is the *event*. The event differs from conventional modes of segmenting time because it doesn't depend on an undefinable *now*. It is not limited to a primal impression, but has a duration that combines particularities, generalities, and living present, and it is defined by a beginning and an end. The living present is not limited to the tick of a clock but is extended by the interplay of the particularities and generalities as they respond to the present object of attention. An event, therefore, is defined by a shift of focus, similar to the way a character's action changes with the circumstances in the scene. You are talking with your friends when the teacher comes into the room—one event ends and another begins. A new set of generalities emerge about the question you have just been asked that replace the plans you were making for the weekend. New particularities are brought into focus and recalling the last class or the reading for the day takes precedence. The seconds pass, but the event changes only when our circumstances do.

The shape of the event occurs through a *contraction*, or the narrowing of focus to the matters at hand. "Events are produced in a chaos, in a chaotic multiplicity, but only under the condition that a sort of screen intervenes." The screen can be seen as a kind of filter that admits to the circle of attention only the relevant information, whether perceptions, memories, or expectations. Deleuze views the world as inherently chaotic, but his use of chaos is metaphorical: "Chaos does not exist; it is an abstraction because it is inseparable from a screen that makes something—something rather than nothing—emerge from it" (Deleuze 1992: 76, 77). The combination of particularities and generalities strives to render the situation intelligible so that we can act appropriately based not only on the influx of information but also on our predispositions. Chaos is avoided through a contraction or focusing in on what is happening now to the exclusion of that which appears to be irrelevant. What emerges through the contraction is the event, our experience of time.

Philosopher Andy Clark approaches these events in a slightly different way. In *Surfing Uncertainty: Prediction, Action and the Embodied Mind*, he explores the science behind 4-E Cognition, or how we engage with the world in which we are embedded. He is less interested in inputs from the environment (although they are significant) than in the emergence of action. As his title indicates, he perceives the world as full of uncertainties (Deleuze's chaos) that have the potential to confound us. Evolutionarily, we would not have survived as a species if we were not able to assess situations and respond appropriately. Key to his theory is our ability to predict what our next actions need to be (Deleuze's generalities and Husserl's protention). Stage managers know when to call the next cue and are prepared to act when the time comes.

Clark believes that we make predictions about what comes next, consciously or not, and that predisposes us to focus on specific perceptual information. "The ceaseless anticipatory buzz of downwards-flowing neural prediction that drives perception and action in a circular causal flow. Incoming sensory information is just one further factor perturbing those restless pro-active seas" (Clark 2015: 52). The data we pay heed to is not necessarily what is most significant, but it is what we predict will be most important. Needless to say, these predictions are not always correct. The stage manager is ready to bring in a sound cue until the actor goes up on a line. There needs to be a quick adjustment to accommodate this disruption. Clark defines ***prediction errors*** as those times when our expectations are at odds with the world. "But should things fail to fall into place (should the results of the perceptual 'experiment' appear to falsify the hypothesis) those error signals can be used to recruit a different hypothesis" (70). We modify our actions to respond more accurately to the circumstances. The result in extreme situations, such as traumatic experiences, may lead to radical "revisions in our model of the world" (80). Nevertheless, the particularities and incoming information lead us to predict how we will behave, rightly or wrongly.

Thinking of events rather than seconds provides a perspective that is in keeping with our experience of time. Unless we are timing a race or an egg on the boil, we are generally unconscious of the clock ticking. Even in something as time-based as music, we are aware of the duration of notes and the regularity of tempo, not the passage of time. Generally, we are conscious of the hands moving on the clock only when extremely bored or anxious for a new event to begin. Phenomenologically, we are more concerned with what is happening around us than time. Disturbances bring an end to this flow. Attention is suddenly focused on the unexpected, and we become more aware of the world around us and the passage of time. Both the flow and its disruption are elements of temporality, inside and outside of the theatre.

Gertrude Stein and time

This chapter began by identifying different registers of time from the universe of Einstein to the phenomenological experiences of everyday life. As we bring our attention back to performance, two additional types of temporality need to be explored. In 1925–6, Gertrude Stein, an American essayist and playwright, wrote a lecture entitled "Composition as Explanation" (Stein 1993: 495–503). Her goal is to evaluate the importance of *composition* in writing and the arts more generally, with particular emphasis on changes in artistic forms from one generation to the next. She defines composition in her enigmatic style:

There is singularly nothing that makes a difference a difference …
except that each generation has something different at which they
are all looking … composition is the difference which makes each
and all of them then different from other generations and this is
what makes everything different otherwise they are all alike and
everybody knows it because everybody says it. (Stein 1993: 495)

Each generation of theatregoers expects something different. Henrik
Ibsen made a break with the traditions of melodrama and the well-
made play when he began to write about his society in what became
known as Realism. That rapidly became what "they are all looking at."
After the Second World War, European playwrights found the realist
form incommensurate with their experience and searched for new
forms that could express their existential dilemma over the Holocaust
and the dropping of the atomic bomb, to say nothing of the deprivation
and destruction that were the fallout of the conflict. Martin Esslin
famously dubbed the plays of Genet, Beckett, Ionesco, and others the
Theatre of the Absurd. That became what avid theatregoers of the
time came to expect.

Stein goes on to define two aspects of composition in relation
to time: "It is understood by this that everything is the same except
composition and time, composition and the time of the composition
and the time in the composition" (Stein 1993: 497). The *time of the
composition* refers to the context in which the work of art was created
and the influence that period in history had on the artist, for instance,
the shift from realism to the absurd. The *time in the composition* is the
actual creation of the work, the effort and how long it took to make it.
In the theatre, these can be further subdivided. In production, the team
not only explores the time when the play was written, but also focuses
on what the play means in the context of today. A dramaturg begins
working on a play by analyzing the text and also investigating the life

Task 1

Think of a theatre performance that you have seen recently, preferably of a classic text. Discuss with your group how the production was interpreted to make it more relevant to the contemporary scene. If you haven't seen one that resonates with your life, what could have been done by the production team to make it more meaningful to you and those around you?

of the playwright and the time in which it was written. These satisfy the two aspects of time and composition posited by Stein. However, she is also looking for resonances the play has with the contemporary world and how to produce it for today's audience. Spectators experience a production in the context of their own time, regardless of when it was written, and they spend time in the composition as it unfolds before them. We will return to a discussion of the audience when we explore the question of aesthetics in Chapter 6.

In the following sections we will look at different productions through the lens of Stein's theory of composition. Rather than focusing on spatial relations, although these will come into play, the emphasis is on the temporal dimension. The choices of productions are performances I have seen, and will therefore represent my experiences, but not necessarily those of others who saw them.

Case studies: *Mise en scène*

Theatre is, in whatever form, a temporal art, and through the changes that take place during the event, the significance of the design elements can change. Design is more than a spatial arrangement with the only changes occurring when a new location is needed. When the set doesn't change, we tend to think of it as a background against which the performance takes place, rather than a vital element in communicating to the audience. Good designs, and in this I include lighting, sound, properties, projections, and costumes as well as sets, provide patterns that mark the passage of time and shifts in mood. What seems to be a happy home in Ibsen's *A Doll House* becomes a claustrophobic hell of betrayal for Nora. The room does not change, but if the designers have done their job correctly the transformation will take place even if the audience is not consciously aware of it. Designs are living organisms that work with the director and actors to tell the story. While this is the challenge faced by all scenographers, I will focus on four productions where the *mise en scène* affected the experience of time.

Einstein on the Beach is a collaboration between Robert Wilson, Phillip Glass, and Lucinda Childs (Obenhaus 1985).[1] It was first created in 1976 and has been revived several times, including in 1992 when I saw it. The Gilman Opera House at the Brooklyn Academy of Music seats around 2,000 spectators in a grand proscenium theatre built at the beginning of the twentieth century. The Phillip Glass Ensemble performs in the orchestra pit while the action takes place on stage. A violinist, made up to look like Einstein, sits down stage right. Behind him are risers on which the chorus stands. Center stage and stage left are open spaces where the various scenes and "knee plays" are staged. The performance test is adapted in part from the writings of Christopher Knowles, a cognitively challenged thirteen-year-old, with additional texts by Lucinda Childs and Samuel M. Johnson.

---

[1]Videos of the performance are currently available on YouTube. I strongly suggest watching them, including the documentary *Einstein on the Beach, The Changing Image of Opera* (https://www.youtube.com/watch?v=HCIrNDgrQLE). Words cannot do justice to complexity and visual impact of the work.

In one of the knee plays, a two-dimensional bus appears upstage left and moves across the stage at a glacial pace, with the rest of the stage in virtual darkness. While the bus moves, the violinist portraying Einstein plays a solo that is accompanied by the women in the chorus. They sing a simple "do re mi," but in the hands of Philip Glass it becomes a haunting adagio. A performer, lit from below as if by the light from the dashboard, "drives" the bus, reciting the following.

> And what sort of story shall we hear? Ah, it will be a familiar story, a story that is so very, very old, and yet it is so new. It is the old, old story of love.
>
> Two lovers sat on a park bench, with their bodies touching each other, holding hands in the moonlight. There was silence between them. So profound was their love for each other, they needed no words to express it. And so, they sat in silence, on a park bench, with their bodies touching, holding hands in the moonlight.
>
> Finally, she spoke. "Do you love me, John?" she asked. "You know I love you, darling," he replied. "I love you more than tongue can tell. You are the light of my life, my sun, moon and stars. You are my everything. Without you I have no reason for being." (Johnson 1976)

The power of the scene lay in the combination of the visual image, music, lighting, and simplicity of the recitation. It was an ethereal experience on its own but more so when juxtaposed with the rest of the production, which included elevators moving vertically and horizontally with performers moving in slow motion, actors in front of a grid of rotating lights pretending to operate a machine, and some dancers doing pedestrian activities while another moved quickly across the stage in abstract bursts of energy.

The production lasted four and a half hours without intermission. While Wilson made it clear that the audience could get up and get refreshments whenever they wanted, on the night that I saw it, everyone watched with rapt attention. The biggest surprise came after the show. On leaving the Brooklyn Academy of Music, I was suddenly assaulted by the noise and energy of the New York streets. The production had altered my perceptual rhythms. It was as if I was brought out of a deep meditative state. My sense of time and the speed of life were temporarily changed, only to be restored abruptly on leaving the theatre. The impact was unforgettable.

In 2008, Katie Mitchell's *Waves,* an adaptation of Virginia Woolf's novel of the same name developed with the National Theatre of Great Britain, was brought to the Duke in New York, a 200-seat black box theatre. Working with videographer Leo Warner, with designs by Vicki Mortimer, lights by Paule Constable, and music by Paul Clark, Mitchell and the company do not enact the book, but choose specific scenes to theatricalize. Woolf's novel is impressionistic, with the environment taking precedence over the narrative events. A haunting work, it reads almost like a memoir but not of the author. While the story is continuous, it spans many years with episodes divided by images of the sea washing over the shore. As the mood becomes less idyllic, the descriptions of the waves become less peaceful, reflecting the tensions between the characters.

The company selected events from the novel, arranged them in sequences defined by simple everyday activities, and then divided these into the discrete tasks needed to convey the story to the audience. When acted, these tasks are assembled into live video feeds projected onto the central screen (see Chapter 3 for a discussion of the spatial design). The audience simultaneously sees the generated audiovisual images on the large screen and the actors creating the video. The remainder of the production consists of the actors (re)creating complex images in plain sight of the audience, who must put the pieces together in much the same way the projections and sound are synchronized.

Neville and Rhoda walk along a rainy street while a narrator reads passages from Woolf's novel about the event. The actor narrating the event stands at the microphone stage right, while other actors spray water on the floor stage left giving the impression of wet pavement. Another holds one of the desk lamps creating the illusion of streetlights shining on sidewalk. A third actor, filming the event, keeps a tight focus on the feet as two others walk, stop, and then continue on their way "down the street." Yet another actor, with two lamps, shines light across the scene to suggest a passing car. In another scene, the actor portraying Jinny looks into a mirror, thinking about an absent Louis. As she looks, he appears in the mirror, a ghostly memory looking over her shoulder. One actor holds a two-way mirror while another lights it from the front, creating the reflection of the young woman. At the correct moment in the narration, which is being read into a microphone at the table, the actor playing Louis is lit behind the mirror, making visible the memory of which she is speaking. As projected on the screen the illusion is very precise.

The spectator has the option of watching the creation of the scene, viewing it projected on the screen, or shifting attention from one to the other. Initially, the viewer feels a degree of distress, not knowing where to look. While the video provides a typical cinematic experience, and therefore some comfort, creating the video in front of the audience is so interesting and ingenious that it is hard not to watch the actors work. This disjunction, not knowing where to look, remained a pleasant challenge for most of the two and a half hours. Finally, without realizing it was happening, it became possible to shift from one to the other effortlessly.

The impressionistic nature of the story and the fragmented structure of the performance may seem alienating and not conducive to strong emotions. However, as the show proceeded, and I became accustomed to the way in which images are put together, the emotional trajectory of the performances grained intensity. By the end, the feelings were quite powerful and unexpected. The emotions were always present but there was no time to appreciate them fully because of the shifting, visual attention. Suddenly, I felt an overwhelming sadness, less at the death of Percival than the loss of youth and the distance that weakens the intimacy of their friendships. That there were feelings is in part due to the narrative and the beauty of Woolf's use of language, but there was also something in the precision and continual activity (although seemingly effortless) in creating the sequence of images that provided a counterpoint underscoring the sadness and sense of loss.

Time plays across the production in several registers. There is the duration of the production, the lifetimes of the characters as they move from childhood to middle age, the creation of the scenes, and the shifting attention from the work of the actors to the image on the screen. While the effect of *Einstein on the Beach* didn't happen until leaving the theatre, the impact in *Waves* is experienced as the production draws to a close, although the overall effect of the production lasted well beyond the moment. Indeed, I still think about and relive the impact ten years later. This is true of all these productions. If the production is good, regardless of form, time is extended beyond the hours spent in the theatre.

Time also was an important aspect of Théâtre du Soleil's *Les Éphémères (The Ephemeries)*. The production was divided into four parts, with each lasting approximately three-and-a-half hours. Brought to New York in 2009, it took place in the Park Avenue Armory in a space designed in the nineteenth century for military drills. The stage ran the width of the room with curtains and false prosceniums at each end. The audience looked at each other across the acting area, sitting in steeply raked tiers of seats. Presented in French, supertitles were projected on a wall above the heads of those sitting in the second row. Ariane Mnouchkine, artistic director of the company, collaborated with composer Jean-Jacques Lemêtre, with sets by Everest Canto de Montserrat; lighting by Elsa Revol and Hugo Mercierm; sound by Yann Lemêtre, Virginie Le Coënt, and Thérèse Spirli; and costumes by Nathalie Thomas, Marie-Hélène Bouvet, Annie Tran, Cécile Gacon, and Chloé Bucas. As the size of the production team suggests, Mnouchkine believes in an open creative process that values the input of numerous collaborators.

The narrative consists of several plot lines that are all linked, some quite loosely, by characters that appear in more than one story. The action begins just after the death of the main character's mother and her decision to sell the house. She thinks there will be time to sort through her mother's things before it is sold, but the house is purchased almost as soon as the for-sale sign is posted. Just beginning the grieving process, a box surfaces containing a few mementoes of her mother's early life. The rest of her storyline consists of following the trail of clues provided by each item. Over the fourteen hours, you come to identify with her search, feeling the depression she experiences. She discovers that her parents protected Jews during the Second World War and came close to being discovered by the Nazis. The play ends with her on the phone, hearing one of the people who knew her mother during the war tell how the Nazis were foiled by an act of bravery and cleverness. Hearing this story, she starts to laugh, a deep and joyous laughter. It was clear that through this detective work, she was able to work through her grief and could continue her life accepting the loss and with a new appreciation for her mother.

Ephemeral means temporary, lasting only a very short time, although it also has the connotation of being immaterial, like gossamer. One can think of life as ephemeral, as in each scene in *Les Éphémères*. Episodic in structure with multiple story lines, the scenes shifted from place to place, seemingly without effort. In fact, there was a great deal of effort. Each scene took place on at least one small platform that emerged from one side of the stage, pushed by stagehands. Some scenes used more than one platform. For instance, when a man bought the house, he stepped onto a small platform that had the for-sale sign before crossing to another with the door on which he knocked. A third platform offered the interior of the house. But most of the scenes took place on one platform, pushed by two stagehands dressed in black. They slowly rotated the platform throughout the scene, either staying in place or slowly moving the length of the stage.[2] The actors played the story without adjusting for the rotations, focusing on the other characters or their particular action. The transitory quality implicit in the title was made manifest through the design with each rotation.

I hope these examples give you an idea of how time is an element of design. I have focused less on costume, lighting, video, and

---

Task 2

Return to the descriptions of the productions in Chapter 2. How did those productions utilize time through design? Focus less on the actors and spectators, and more on how the use of space also affected the experience of time.

---

[2]There is a short video on YouTube of the movement of the platforms. Viewed April 9, 2018. https://www.youtube.com/watch?v=SmRK7As7Vcs.

sound. Each of these also affects the spectators' experience of time in performance. For instance, in a recent production of *King Lear*, Cordelia began the play dressed in white, but when she returns from France with her army, she was in deep burgundy and armor. At her death, however, she was once again in white. The changes in costume accentuated the passage of time while also keying the audience into who she has become through the events in her life. In a production of *Apparition: An Uneasy Play of the Underknown* by Anne Washburn, Mallory Catlett used curtain hampers as set pieces in this exploration of unconscious fears through the lens of *Macbeth*. Through arranging the hampers, she created a forest, a hallway, a bridge rampart, or simply obstacles to be negotiated. The use of light accentuated the mood, refocusing the attention from the audience to different parts of the theatre, including the catwalk surrounding the stage, to mark changes of location. In *This Was the End*, Catlett projected a video of the set onto the set, in a 1:1 ratio. By slightly increasing and decreasing the aspect ratio of the video, the set seemed to pulse, underscoring the motif of trauma in the production. Later, characters were projected onto the design with the live actors playing the same characters present on stage, emphasizing the fragmentation experienced by victims of traumatic experiences.

Task 3

Recall a performance you have seen. How did the design help to define the passing of time? Think about the passage of time in the narrative, and in supporting the meanings you derived from the experience.

Playing with time

The French director and actor Jean Louis Barrault writes that a play is "interrupted silence" (Hodge and McClain 2010: 6). As the curtain rises, and a hush of anticipation falls over the audience, there is a silence that is only broken when the first words are spoken. "I remain quiet. I don't breathe; none of us breathes; we vibrate in immobility; and we discover that unique *Silence* which alone can give us the physical sensation of the *Present*" (Barrault 1951: 16). He believes that it is in the silences, framed by language, that we experience the present not as an absence but as a vibration. Constantin Stanislavski echoes Barrault: "I was silent and you made the effort to penetrate the meaning of my silence" (Stanislavski 1949: 136). There is a difference, however, between them. The French director focuses on the communication between the performer and audience, while Stanislavski emphasizes the spectator's effort to discover the interiority of the character.

As Barrault rightly claims, silences have always been part of the theatre from the Greeks to the present. However, Anton Chekhov was the first to make them a significant part of the action, providing performers the opportunity to investigate the character's uncertainties for the audience. They became a more central part of modern dramaturgy with Harold Pinter and the "Pinter pause." With Pinter the pause was accompanied by silences, offering longer periods of time for the audience to explore the tensions on stage without the help of language. The trend started by Pinter lasts to the present day, perhaps showing a distrust of language or the realization that we spend much of our time without speaking. We will discuss Pinter and other playwrights who use pauses in the next chapter on language. Here we will look at performances that use tempo to mark the experience in visceral ways after we examine the concept of silence.

On August 29, 1952, John Cage premiered *4'33"*, a work for piano during which not one note is played. The music, as envisioned by Cage, consisted of the sounds made by the audience, and the ambient noises in the concert hall. While the composition reflects the composer's interest in chance and Zen philosophy, it also challenges the concept of silence. Cage recounts a more direct experience that took place at Harvard where he entered an anechoic chamber:

> Its six walls made of special material, a room without echoes. I entered ... and heard two sounds, one high and one low. When I described them to the engineer in charge, he informed me that the high one was my nervous system in operation, the low one my blood in circulation. Until I die there will be sounds. And they will continue following my death. One need not fear about the future of music. (Cage 1961: 8)

Silence signifies an ideal.

Samuel Beckett comes to the same conclusion in *Not I*, a monologue spoken by a mouth and heard by an onstage auditor. The narrator tells the story of an abused and ostracized woman; insisting that the tale is not autobiographical, the woman is "not I." In relating the story of a pariah, the mouth attributes to her the inability to use language, and when she speaks it is an incoherent stream of sounds: "Words were coming ... a voice she did not recognize ... at first ... so long since it had sounded ... then finally had to admit ... could be none other ... than her own ... certain vowel sounds ... she had never heard ... elsewhere." Those times when she is not speaking, "practically speechless ... all her days," there was no silence, "the buzzing? ... yes ... all the time the buzzing ... dull roar ... in the skull" (Beckett 1984: 218–19, 221). The persistent sound of the nerves proves against all desire that she is still alive. The only silence is when the buzzing stops.

Ariane Mnouchkine's production of *Les Atrides* consists of four Greek plays: Euripides' *Iphegenia in Aulis* and Aeschylus' trilogy *The Oresteia*. The show was presented in the Park Slope Armory in Brooklyn over two days, in ten hours, with two of the plays performed each day with a meal break between them. As is customary with the Théâtre du Soleil, the audience could walk behind the seating and watch the performers preparing for the performance. While I did not expect this, I was more struck by a constant drone that filled the armory with ambient sound. It established an ominous sense of foreboding. The set was painted desert sand and resembled a square bull-fighting ring. Upstage center was a wide gate, and stage left was a two-story structure on top of which Jean-Jacques Lemetre performed the music that accompanied the production.

During the dialogic scenes, the actors struck stylized poses they held virtually without moving, reciting the verse quickly with a minimum of inflection but with great intensity, reminiscent of what Vsevolod Meyerhold called an "internal mystic vibration" (Braun 1969: 54) and a *"plasticity which does not correspond to the words"* (56). The stillness of the acting during the episodes contrasted radically with the movement of the chorus, which entered through the upstage gates, choreographed to swirl in multicolored costumes and stamping in unison that made for a powerful presence. The chorus did not speak; only the choral leader interacted with house of Agamemnon.

The alternation between the powerful reserve of the royal family and the majestic dances of the chorus created a pattern that was broken by the appearance of the corpses, first of Agamemnon and then Clytemnestra and Aegisthus. A particularly powerful moment was during *The Libation Bearers*. Electra and Orestes, with the aid of the chorus, enacted a ritual to bring Agamemnon's body from the grave. Electra led the chorus in a ritualized cadence that grew in rhythmic intensity as the lighting became darker and darker. It was clearly a female ritual as the chorus of women became possessed by the call to the dead. Orestes was excluded from the act, afraid and in awe of the power of the women. The scene worked to a crescendo with a sudden cessation and the appearance of the body of the Agathos. The play ended with the furies accepted into Athens as protectors of the realm. Unlike traditional choruses, in Mnouchkine's production they appeared in steel gray and black dog costumes, barking and snarling at the audience in canine fury. The resolution to the matricide, far from bringing a sense of peace restored, was full of foreboding about an anguished and violent future.

The quality of the performance by the actors, the beauty of the *mise en scène,* and Lemetre creating a soundscape in a forest of instruments were mesmerizing. Each two-and-a-half-hour segment of the production created a desire for more and passed seemingly in a flash. As in *Einstein on the Beach,* discussed earlier, the experience of time was altered through the changing rhythms and dynamic intensity of the performers, some still, others in martial movement creating a world pulsing with life and regal power in plays about death and retribution.

The desire was for the play to continue beyond its limits. In some ways it did. I still think about the production and feel the wonder and passion evoked in me during those two days of performance. It continues to live in imperfect memory through brief images that return me to the armory and that vision of Greek tragedy.

Performer and audience

We never go to the theatre without a set of expectations. There is a reason we go to a particular production. Our anticipations are based on the particularities of past experiences with theatre and performance, what we know about the production, and whatever details we know about the actors and/or the company. These expectations are generalities bringing together past events, current circumstances, and predictions about what is to come. These give rise to the excitement we feel when entering the theatre.

Going into the theatre can affect your feelings about the performance. When I entered the Golden Theater to see the Donmar Warehouse Production of John Logan's *Red*, Alfred Molina, who played Mark Rothko, was already sitting on a chair down stage center with his back to the audience. "The set of his neck and shoulders makes it clear that he is staring hard and hungrily, locked in visual communion with the object before him," a copy of one of Rothko's paintings (Brantley 2010). *Falls for Jodie* is Eric Micha Holmes' play about John Hinkley's obsession with Jodie Foster and his attempt to assassinate Ronald Reagan. Entering the theatre, I was told to proceed to a set of stairs leading onto the stage. Climbing the steps, I found myself on the edge of a hotel room. The set was surrounded by a small number of chairs designated for the spectators, who were a hair's breadth of where the play would take place. Both openings intrigued me but established very different expectations about what I was about to see.

Every spectator has different expectations because they have a different history of going to the theatre, other cultural events, and life experiences. When the performance begins, these anticipations recede and, if the performance is worth the price of admission (even if it is free), we become immersed in what is happening on stage. As we watch the show, we are never passive, but always actively engaged, perceiving and responding, enjoying and judging, feeling and reflecting on what the performance asks of us.

Actors want an audience, and they want them to be present, journeying through the production with them. Their presence is tangible. They know when the spectators are engaged in the performance and when they are restive. They can tell by the laughs, the coughs, and when all is working well, the stillness in the room. Depending on their experience, the viewers can tell if the actors are working hard or "telephoning" it in. By working hard, I don't mean a lot of apparent effort, because a quality performance seems effortless despite the amount of "extra daily energy," to borrow a phrase from Eugenio Barba, they are burning up. The performers and the audience need to collaborate with each other. It doesn't always happen but when it does it is magical.

I avoid using the word "relationship" in discussing the interaction between audience and performer. Instead, I prefer *correspondence* thanks to Tim Ingold who, in *The Life of Lines*, defines correspondence as "how knowledge grows from the crucible of lives lived with others ... This knowledge consists not in propositions about the world but in the skills of perception and sensuous engagements with the beings and things with whom, and with which, we share our lives" (Ingold 2015: 1). We come to understand what is taking place before us through the intense engagement of actors and spectators during the time of the event. Performers and spectators may be separated by the distance between the stage and the house but, when theatre works, space gives way to the dynamics of time and the sensory engagement commitment of the audience.

Ingold uses a metaphor to explain this interaction. He challenges the concept of something happening "between" people and chooses the image of a river to do so. To define a river by its banks is to think of it as existing *between*. "'Between' articulates a divided world that is already carved at the joints. It is a bridge, a hinge, a connection, an attraction of opposites, a link in a chain, a doubled-headed arrow that points at once to this and that" (Ingold 2015: 146). Such acts of differentiation are the result of analysis, and they are one way we learn. He identifies another way of gaining knowledge: *in-between*.

"'In-between', by contrast, is a movement of *generation and dissolution* in a world of becoming where things are not yet given ... Between has two terminals, in-between has none" (Ingold 2015: 147, emphasis added). Ingold believes that gaining knowledge is inherent in our engagement with the world. Between marks a conclusion reached, like the banks defining the river. In-between is where the river flows, where experience generates knowledge.

In using correspondence, I wish to stop thinking of the relationship *between* actor and audience and replace it with "a movement of generation and dissolution in a world of becoming where things are not yet given." Watching the production, our attention is drawn to the events that are taking place on stage for the first time, at least for the spectators. We come to an understanding by following the action occurring on the stage and in the context of the world of the play. The *mise en scène* creates an ecology in which we become embedded as embodied beings. The correspondence that develops between the production and the spectator occurs in time and takes place in the distance that separates the audience and the performers.

A particularly rich example of the correspondence between performer, design, and time is *The Concept of the Face: Regarding the Son of God,* a production by the Italian director Romeo Castellucci and the Societas Raffaelo Sanzio. The performance takes place in three sections. In the first, the stage setting is divided into three living

Task 6

Think of a performance that worked for you. Try to remember how the connection between you and performer/production felt. What was it about the acting that engaged you? How would you describe your engagement with the actors?

areas furnished in white and chrome. Stage right is a living room, center stage a kitchen table and chairs, and stage left a bedroom. At the back of the stage is a floor-to-flies rendering of the *Salvator Mundi* by Antonello da Messina (1465–75), depicting the *Ecce Homo,* or Christ after being scourged during the passion. Despite the suffering he endured, there is no sense of pain, condemnation, or forgiveness— the face is simply passive. The first part of the performance is the narrative of a middle-aged man taking care of his old and incontinent father. In the course of the action, they move from the living room to the bedroom, and in each of the three areas, the father soils himself, each one more violent and graphic than before. The first event is telegraphed by a brown stain on the back of his white robe while the last by what can best be described as projectile diarrhea. Each time the son cleans his father; each time he cares for his father with love and compassion. The father is increasingly humiliated by his inability to control his bowels, while the son grows increasingly frustrated by the repeated need to wash his father.

The graphic realism of the feces and the subtle smell pumped into the theatre made me look away in disgust. Whenever I did, my gaze focused on the painting at the back of the stage. After the initial shock, I looked back at the characters and was drawn into the touching interaction between the two men. As the pattern was established, I became increasingly aware of the difference between the passion of what was happening on stage and the passivity of the painting. French philosopher Emmanuel Levinas has written eloquently on the "face of the other" as presenting an obligation to respond to the need of the other. "The being that expresses itself imposes itself, but does so precisely by appealing to me with its destitution and nudity—its hunger—without my being able to be deaf to that appeal" (Levinas 1969: 200). It was apparent, as the production proceeded, that the painting was "deaf" to the anguish that was taking place on stage.

Over time, both during the show and afterward, the distance between the narrative and the image grew and a critique (accurate or not) of the Catholic Church became evident: it is not responsive to the suffering of others.

I talked earlier of *Sleep No More*, the immersive theatre piece. The surreal effect I experienced resulted from the time spent exploring the space and the sudden appearance of the performers. Their presence and disappearance as the audience in their masks explored the environment created a dream-like experience. It was neither the performances nor the design alone but working in correspondence with the spectators—over time—that the power of the production revealed itself. Every production manipulates time, some more evidently and capably than others, but the event is rhythmically structured in order to create an effect that will enthral the audience. When it works, therein lies the power of theatre.

5

The Text

When we think of theatre, we think of scripts: texts written by a playwright, learned and performed by actors. While this is clearly not the case in all forms of theatre, there is an assumption among most audience members that when they go to a show they are going to see a play. In this chapter we will look at language, used both in everyday life and in performance. In the process, we will encounter concepts such as structuralism, cognitive blending, metaphor, and crossing. In addition, we will look at language as embodied speech and the intersection of speaking and gesture. This entails a look at the differences between spontaneous speech and that spoken from memory. Finally, we will look at performances that do not use spoken language.

In the beginning there may have been the word, but before the word there was silence. While much is communicated in the theatre through the spoken word, just as much information, and often the most interesting, is in the pauses. Whether it is Iphigenia's realization that she is to be sacrificed so the Greek fleet can sail to fight the Trojan War, or the pregnant pauses associated with the works of Harold Pinter, the moments of silence can speak volumes. But the quiet is noticeable only because of the words that frame it. We notice stillness because we have just left a noisy room. In everyday life the silence can seem a relief, but in the theatre, it piques our curiosity: we wonder why they have stopped talking. In the plays of Pinter, we sense a tension between the characters that bridges the pause and asks us to ascertain what the characters are thinking and feeling. But that only happens because of the words.

Language

Language is learned through an ability to imitate, as we match vocalizations to the sounds that we hear. My granddaughter, as many before and after her, is learning to speak and, as she progresses, becomes better able to match the sounds she makes to those she perceives. She still says "nana" for "banana," for instance, and "sarkle" for "sparkle." Similarly, the Pimsleur approach to learning a foreign language is based on saying aloud the words that you hear and consolidating them in memory through increasing intervals between repetitions. The key is connecting the sound of the words with the musculature needed to produce the sounds. Language, in short, is embodied and depends on our ability to perceive accurately and control the vocal mechanism in the production of sounds. This anecdotal evidence is supported by psychologists working in the UK: "Consistent with this position, the results reviewed here suggest that brain regions and networks involved in speech production are ubiquitously involved in speech perception" (Skipper, Devin, and Lametti 2017: 97). This may also apply to reading. Although the sample is small, many people I have talked to hear the words as they read them. Others, though, like my wife, don't need to hear them to comprehend what is written.

Scientists discovered a phenomenon known as mirror neurons while studying macaque monkeys. What scientists discovered is that the neurons that fire when the animal undergoes a task such as picking up an object are also activated when they see another monkey or human doing the same activity. They fire at a lesser intensity but the same synapses are energized. Italian neuroscientist Vittorio Gallese, one of the discoverers of mirror neurons, hypothesizes that these neurons may help us understand language acquisition as a social event. "To imbue words with meaning requires a fusion between the articulated sound of words and the [communally] shared meaning of the experience of action" (Gallese 2008: 329). While there is significant controversy about the function of mirror neurons, his theory provides an evolutionary approach to the origins of language that avoids the emergence of a new faculty or innate grammar in the human brain, a theory proposed by Noam Chomsky and others (Everett 2017).

This does not mean that other ways of thinking about language are not without merit. Ferdinand de Saussure, one of the founders of **Structuralism**, investigated the relationship between words and what they signify. **Sign** is the term he used in developing his theory. His work in linguistics led him to conclude that a sign is made up of two aspects, the **signifier** and the **signified**. The first is the word used in the place of the latter, that to which it refers. The technology that I am using to write (the signified) is associated with the term 'computer' (the signifier). Charles Sanders Peirce was working at about the same time as Saussure, but independently of him. He also identified the sign, signified, and signifier, but believed there was a fourth term: the **interpretant**. The interpretant is the class of concepts/objects related to the signified. For instance, the word table does not refer to one ideal form of table, but to the whole class of tables with which we are familiar. This addition opens the sign to a wide range of possible meanings. If you think of the words love or nationalism, you do not think of a single example, but a whole range of experiences and meanings. The shift is the difference between denotative and connotative meanings of words. One is constrained, a virtue in communicating clearly, while the other is expansive, allowing for more poetic uses of language.

Language allows for both clarity and specificity, as well as flights of imagination. In the theatre both are used. Information about characters, given circumstances, backstory, and location all benefit from the specificity of language. If that is all there is to plays they would read like news stories. What makes theatre exciting is the use of poetic language, no matter how prosaic (such as in the works of David Mamet), that surprises the spectator by introducing the unexpected, whether it is a verbal attack or a moving use of language: "If music be the food of love, play on." In this chapter, we look at metaphorical uses of language and how certain playwrights employ language and silence in the structuring of their plays.

Metaphors and conceptual blending

Cognitive linguist George Lakoff and philosopher Mark Johnson write that "The essence of a metaphor is understanding and experiencing one kind of thing in terms of another" (Lakoff and Johnson 2003: 5). They go on to say, "In actuality we feel that no metaphor can ever be comprehended or even adequately represented independent of its experiential basis" (Lakoff and Johnson 2003: 19). A metaphor, they argue, is not merely a turn of phrase, but depends on our formative experiences as sentient beings. They have elaborated this theory in *Metaphors We Live By* (2003) and *Philosophy in the Flesh: The Embodied Mind and Its Challenge to Western Philosophy* (1999). Both books develop a ***theory of cognitive metaphors***, exploring how the tropes we use every day can be traced back to our earliest experiences. "There is no such fully autonomous faculty of reason separate from and independent of bodily capacities such as perception and movement. The evidence supports, instead, an evolutionary view, in which reason uses and grows out of such bodily capacities" (Lakoff and Johnson 1999: 17).

In the process of examining the relationship between metaphors and experience, they develop a taxonomy based on differences and similarities between different tropes. Some categories they have developed include CONTAINMENT ("I am in a relationship"), LOVE IS MADNESS ("I am crazy for you"), and SOURCE-PATH-GOAL ("life is a journey"), to name but three. They also distinguish between primary (ORIENTATION) and secondary (HEAT IS UP, COLD IS DOWN) metaphors, the latter defined as subsets of the former. It is also possible to blend different metaphors ("being in love is hot"). It is not possible to do justice to their work in one chapter, so perhaps one example can give an idea of the correspondence between the body and metaphors. We experience temperature changes throughout our lives. We act to preserve heat when it is cold by making ourselves compact, hugging ourselves, or putting our hands in pockets. When it is warm, we open up to take up more space in order to dissipate heat. Lakoff and Johnson see these as the basis for metaphors, such as "the economy is heating up." It is seen to be expanding with an increase in sales and growth in the gross national product, which we feel is a positive thing. When there is a downturn, we say "the economy is cooling off." In actuality, the economy is getting neither warmer nor colder, but we use those terms because they make sense based on our physical experience of hot and cold. They are, according to the definition, metaphors.

Task 1

In small groups or individually make a list of phrases that you use regularly every day. How many of them are metaphors? What kinds of physical experience are they based on, such as containment, source-path-goal, or another category. If you have trouble thinking of them, focus on conversations you have with your friends, or that come up in your courses.

There are critiques of Lakoff and Johnson's work. Some question if certain examples used by the authors are really metaphors or more accurately conceptual descriptions. Another criticism is voiced by Stephen Pinker: "The methodical use of metaphor in science shows that metaphor is a way of adapting language to reality, not the other way around, and that it can capture genuine laws in the world, not just project comfortable images onto it" (Turney 2009). None of these criticisms, which focus on presentation and lack of empirical evidence, actually call into question the validity of the theory of cognitive metaphor. Rather, George Lakoff is accused of overstating the case in his enthusiasm for the work he has done on embodied cognition.

While Lakoff and Johnson describe how metaphors are derived from bodily experiences, Gilles Fauconnier and Mark Turner note that "some things are not metaphors; instead, they are an integration network of two or more concepts or 'mental spaces'" (Cook 2018: 20). They call their theory *conceptual blending* or *conceptual integration* (see Figure 7). Their approach is complex, and only the basics can be

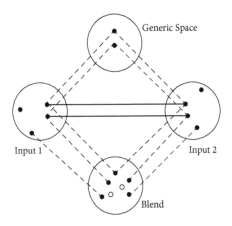

Figure 7 *Fauconnier and Turner's diagram of a simple cognitive blend (Fauconnier and Turner 2002).*

addressed here, as with the discussion of embodied metaphors. The theory seeks to understand how we take disparate concepts and bring them together into an integrated, imaginative structure of meaning that allows us to see in a new and more complex way. As such, the blend is not a static image but a dynamic and imaginative way of combining experiences and memories with present circumstances to decide on a course of action.

The basic structure of a cognitive blend consists of four parts. There are two ***input spaces*** for the concepts to be blended. Amy Cook of Stony Brook University uses Margaret Brown Wise's *The Runaway Bunny* (1942) as an example. One episode in the book has the little bunny declaring he is going to swim away from home. The mother says that she will become a fisherman who catches and reels the baby back to her arms. The picture in the book shows the older bunny fly fishing with a carrot as bait, using food as a means of luring the youngster home. One input space is occupied by the mother bunny and the other with a fisherman. Each has qualities, both denotative and connotative, that allow us to form a blend. The qualities associated with fishing and mother rabbits are ***entailments***. For example, a bunny is a small, furry creature that is mostly mute that eats vegetation, and so forth. The ***generic space*** contains those elements that the bunny and the fisherman have in common. They are both trying to catch something using a rod and baited reel in hopes of luring their chosen prey to be hooked and landed. There are excluded elements that don't fit into the blend: the fisherman intends to kill and eat his catch while the mother bunny is hooking the baby in order to protect it from harm. Those elements from the input spaces that resonate with each other in the generic space are then integrated in the ***blend***. We accept, in the fictional world of the book, that mother bunnies can fish and use carrots as bait to keep their babies from running away.

The elements of the blend are not autonomous but take place within a context (the baby bunny intends to run away, and the loving mother is determined to keep him at home). The where, when, and why of the story determine which entailments are brought into the blend and those that are excluded. In *The Matrix*, a movie by Lilly and Lana Wachowski, technology has taken over and enslaved human beings, using them as a source of energy. All the positive uses of the digital age are excluded from the blend because they are irrelevant to the story the filmmakers are trying to tell. Instead, reality is seen to be an illusion that keeps people from seeing the truth of their existence. As with *The Matrix*, we accept the images without needing to understand the processes that go into making the blend. Fauconnier and Turner contend that blending is not a conscious act, but reflect how cognition works in everyday life or, in the case of *The Runaway Bunny* and *The Matrix*, in artistic expression: "Nearly all important thinking takes place outside of consciousness and is not available on introspection; the mental feats we think of as the most impressive are trivial compared to everyday capacities; the imagination is always at work in ways that consciousness does not comprehend" (Fauconnier and Turner 2002: 33–4). Blends are not limited to creative acts, although they are certainly of major significance, but are part and parcel of how we think.

Task 2

Select a scene from a movie or theatre production. Pick a particular scene and analyze how the artists created a blend, such as why a particular actor was chosen to play a particular role. Identify the entailments that went into making it.

Conceptual blends are not static accomplishments. Once formed, they can continue to morph as new entailments are brought into the blend and others are moved out of the input space. Fauconnier and Turner call this the **running of the blend**. Our attitudes to the machine in *The Matrix* change as the movie progresses. We learn that what is first accepted as everyday life is actually the creation of artificial intelligence, and we gradually come to understand how the matrix is used to keep the population that is fueling the machines from realizing the paucity of their lives. Like metaphors, blends are powerful cognitive tools that are central to being the people we are.

Crossing

Language is not a separate mental faculty but is inextricably integrated into our worldly lives. Metaphors and blends help us understand experience, but equally important they are foundational in our ongoing engagement with the world. Thus far, the discussion has focused on successful strategies for understanding and communicating. In this section, we will explore what happens when language fails and how we grapple with those moments when we can't find the right words. Philosopher and psychoanalyst Eugene Gendlin writes that "experience must be referred to directly—it must be thought of as that partly unformed stream of feeling that we have every moment. I shall call it 'experiencing,' using that term for the flow of feeling, concretely, to which you can every moment attend inwardly, if you wish" (Gendlin 1962: 3). What happens when that flow is disrupted and we become aware of uncertainty?

Anyone who writes experiences the despair of not knowing what comes next. There are moments when the words flow effortlessly onto the page, but we also experience our mind going blank, even though we think we know what we want to say. Psychologist Rolf Reber calls the flow of experience *fluency*: "Processing fluency, or simply fluency, is defined as the ease with which information flows through the cognitive system (which includes both perceptual and conceptual components)" (Reber 2014: 225). This is similar to the concept of *flow* advanced by Mihaly Csikszentmihályi in his book of the same name: "'Flow' is the way people describe their state of mind when consciousness is harmoniously ordered, and they want to pursue whatever they are doing for its own sake" (Csikszentmihályi 1990: 6). While Csikszentmihályi is identifying a way to happiness, Reber is describing everyday life. When things are going well, we are fluent and words come easily, but he also identifies *disfluency*, when the easy movement of experience is broken and words fail us.

Disfluency occurs when our predictions of what happens next falter, either because what occurs exceeds our expectations or falls short. In the act of writing, it tends to be the latter that takes place, although we can also be surprised by a new way of formulating an idea. Regardless, the loss of flow requires or at least allows for a moment of reflection. We evaluate how best to bring what we want to say into a successful communication, seeking the means of expression that will coalesce an idea that is slowly being formed. Herbert Blau, one of the great theorists of the theatre, was fond of saying, "If I knew what I wanted to say, I would be saying something different" (Blau 1986). He was in part commenting on the inability of language to capture an idea, but he was also arguing that repressed desires keep us from knowing what underlies our thoughts. They find expression only indirectly, concealing what would be disruptive if acknowledged in writing. Whether or not you accept theories of the symbolic and repression, the fact is that writing is not a straightforward process. We constantly need to revise as we try to say as precisely as possible what we want to say.

The phenomena of fluency and disfluency also interest Gendlin, but from a slightly different direction. He conceptualizes fluency and flow as **experiencing**: "What I call 'experiencing' is not separable from concepts, but it plays crucial, directly demonstrable roles in ongoing thinking. It performs functions that concepts cannot perform" (Gendlin 2004: 128). The phrase he uses to characterize ongoing thinking is **carrying forward**. As I write or carry on a conversation my thought process is carried forward. The ideas flow and the words appear that allow me to continue my train of thought. The image of a train is an apt one, as one word follows on the next and the meaning is sufficiently clear to allow for communication; in a way this is reminiscent of Deleuze's categories of time. The particularities of what has just been written correspond to the generalities of what I want to say and what I write. Gendlin calls this movement through time **intricacy**, rejecting as simplistic the metaphor of a train. Language is linear when it appears on the page, but not when it is being formulated. Rather than seeing words leading the process of writing, an array of possibilities offers multiple ways of moving thought forward. He sees disfluency not as a failure of language but as the outcome of an intricate process of selecting the words that fit the idea and the idea that fits the words. "Where others see indeterminacy, we find intricacy—an always unfinished order that cannot be represented, but has to be taken along as we think" (Gendlin 2004: 128). As long as I don't struggle to find the right word, typing flows with relative ease. However, as is often the case, when what I want to say does not come easily, the typing stops and I have to regroup, often going back over what is on the page to reestablish and clarify my thoughts. Relief washes over me when fluency returns.

Gendlin believes that writer's block is only a temporary setback regardless of how persistent and torturous. That is why the cure for the malady is to keep writing every day, because eventually a right

---

Task 3

Write a precise and concise summary of what you have just read. Be aware of those moments when you have to pause to think about what comes next. Be aware of the different options you undertake before you are happy with what you have written.

---

order will materialize and the process can move forward. "But if one recognizes that language is inherently metaphorical and not controlled by concepts, then there need be no dead end" (Gendlin 2004: 127). In other words, it is not a question of writing a predetermined sequence, but of immersing ourselves in the possibilities resonating between words, feeling, and thinking. Writing is not linear but a question of selecting words out of a field of possibilities that allow us to communicate with the greatest clarity. Gendlin calls this process *crossing*, or a blending of intricacies. Different combinations offer alternative ways of expressing the same thought. He does not see this as a problem, arguing that crossing enriches the process, rewarding the search to find the best word. "The crossing of two junctures does not bring the lowest common denominator but rather a great deal that is new to both of the two that cross" (Gendlin 2004: 135). It not only enriches the immediate phrase but also alters what comes before because an additional way of formulating the problem has been brought into the mix of what has already been and is to be written.

Writing for the theatre

Playwrights remain the primary source of theatrical texts. But in the age of post-dramatic theatre, plays are not the only foundations for creating performances. Forced Entertainment brought *Real Magic* to La Mama, ETC in New York in 2017. The production used a quiz show as the basis for exploring the human propensity for remaining optimistic in the most impossible of situations. There were three performers who repeated a short skit over an hour. One actor played the emcee, another the contestant, and someone asked the third to think of three words. The participant was asked to guess what word the other was thinking of. In each repetition, there were the same three words. These were shown to the audience on signs, while keeping the contestant in the dark. They were algebra, caravan, and sausage. Every time the guest gave the same wrong answers. The host would vary the chatter but always with the same result: the contestant failed to guess the correct answers. It wasn't a play in the traditional sense, but a text for performance.

However, in this section, I focus on plays written by twentieth-century playwrights: Samuel Beckett, Harold Pinter, Franz Xaver Kroetz, and Marie Irene Fornes. While all are known for their distinctive use of language, the focus is on their use of pauses, ranging from a short breath to silences thirty seconds in duration. In addition to the speeches, all four are also known for their thoughts about staging and clues for the actor. These can be brief as in the case of Beckett or, in Kroetz's *Request Concert,* there are only stage directions, without a word being spoken.

Samuel Beckett's *Not I*

Not I is a monologue. The narrator, with only her mouth visible, speaks to an auditor, who is shrouded in darkness and shrugs four times in the course of the performance. The narrator tells the story of an abused and ostracized woman who, at seventy, suffers a catastrophic brain injury, probably a stroke: "a few steps then stop ... stare into space … then on … a few more … stop and stare again … so on … drifting around … when suddenly … gradually … all went out" (Beckett 1984: 216). She becomes aware she is still alive because of "the buzzing? … yes … all dead still but for the buzzing […] dull roar … in the skull" (218–19, 221). Similar to Cage's "nervous system in operation," the speaker regains awareness and relates the woman's life history, insisting the tale is not autobiographical (hence the title *Not I*).

The British actor Billie Whitelaw performed *Not I* with Beckett directing (Beckett and Whitelaw 1973). What is awe-inspiring about Whitelaw's performance is the intensity, the pace, and the lack of psychologically based emotions. The pauses break the stream of words abruptly, as if stabbing with a knife, but they also establish a rhythm that is punctuated by the choice of words and the insertion of screams and silences:

> Couldn't make a sound … not any sound … no sound of any kind … no screaming for help for example … should she feel so inclined … scream … [*Screams*] … then listen … [*Silence*] … scream again … [*Screams*] … no … spared that … all silent but the grave … no part—… what? … the buzzing? … yes … all silent but for the buzzing … (218)

The language consists of the "s" alliterations, first with the softness of "sound" and then the sharpness of the "scream" followed by the sibilance of "silence." The use of the s's creates a sense of fluency, while the use of ellipses, the screams, and the shrugs of the auditor create a disfluency. The audience member is brought forward haltingly, caught up in the flow of language but brought up short by the breaks in the narrative.

The ellipses mark breaks in thought that, given the pace of delivery, require the actor to make sudden and sharp shifts in thinking and force the audience to keep pace if the play is to make sense. The screams and the silences provide the spectator with a brief respite, although hardly the ability to relax, before the onslaught of language begins again. The speed of the recitation defers empathy with the character through the end of the play, when as the curtain descends the "*Voices continues behind curtain, unintelligible*" (223). The intentional irony of the play is an onslaught of language spoken, auspiciously, by someone who does not speak. Yet, the voice does want to speak. She struggles throughout the play to formulate, intermingling fluencies with disfluencies, what she is compelled to say. She is finally able to forge: "God is love ... tender mercies ... new every morning" (221–2). Then the voice becomes unintelligible, and she goes on.

Harold Pinter: *No Man's Land*

Many of Pinter's plays, *No Man's Land* being no exception, present a surface reality into which a foreign element is introduced that disrupts the surface of complacency with a menace that drives the action of the play. The dramatic tension is tangibly felt in the pauses as the characters strive to understand the nature of the threat and its potential for perturbing their lives. Two poets, one more famous than the other, spend a night in the home of Hirst, a very successful writer who is taken care of by Foster and Briggs. Apparently, there was an earlier relationship between the two authors, and Spooner would like to live with Hirst as his valet. His efforts are resisted by the two attendants and end with Spooner seemingly admitting defeat: "You are in no man's land. Which never moves, which never changes, which never grows older, but which remains forever, icy and silent" (Pinter 1975: 86).

In the first act of *No Man's Land*, Spooner and Hirst are talking. Spooner, a poet, has appeared, apparently without invitation, in the house of Hirst, a more famous poet. As they share drinks, Spooner asks him about his wife. Getting no response, he questions Hirst's performance as a husband.

> **Spooner** I begin to wonder whether you do in fact truly remember her, whether you did truly love her, truly caressed her, truly did cradle her, truly did husband her, falsely dreamed or did truly adore her. I have seriously questioned these propositions and find them threadbare. *Silence.* Her eyes, I take it were hazel?
>
> **Hirst** *stands, carefully. He moves, with a slight stagger, to the cabinet, pours whisky, drinks.*
>
> **Hirst**   Hazel shit.
>
> **Spooner** Good lord, good lord, do I detect a touch of the maudlin? *Pause.* Hazel shit. I ask myself: Have I ever seen hazel shit? Or hazel eyes, for that matter?
>
> **Hirst** *throws his glass at him, ineffectually. It bounces on the carpet.*
>   Do I detect a touch of the hostile? Do I detect—with respect—a touch of too many glasses of ale followed by the great malt which wounds? Which wounds?
>
> *Silence*
>
> **Hirst**   Tonight … my friend … you find me in the last lap of a race … I had long forgotten to run.
>
> *Pause*
>
> **Spooner** A metaphor. Things are looking up. (Pinter 1975: 20–1)

After Spooner's attack, Hirst is silent. Spooner presses his advantage by identifying the color of her eyes—he may or may not have met his wife. Hirst's memories are apparently not totally positive (his wife may or may not have had an affair, including with Spooner), identifying the color of his wife's eyes as "hazel shit." Spooner persists, intentionally misinterpreting Hirst's interjection as maudlin, which causes Hirst to throw his glass to the floor. Spooner completes his attack by asking Hirst if he is merely drunk. After a silence, Hirst makes a pronouncement that is nostalgic *and* maudlin. Spooner then completes the section with a quip that sustains the humor woven into their dialogue. The comedy serves at least a double purpose. First, it underscores the relentlessness of Spooner's attack on Hirst. Second, it brings the audience out of the silence or pause, returning them to the surface of the action through laughter.

The assumption I am making is that the silences and pauses provide a short period of time for the spectators to imagine what Hirst is going through and, since the dynamics of power shift throughout the play, reassess his relationship with Spooner. Richard Ashley of Northwestern University, in a paper on "Music, Narrative, and Emotion," analyzed Claude Debussy's *Reverie*. The composer added a few beats of silence in the penultimate moments of a variation. Ashley argues that this brief period of time gives the listener an opportunity to imagine how it would be resolved. Then the phrase is concluded as the composer wished (Ashley 2012). Like the music prepares the listener of *Reverie*, Pinter and the actors prepare the audience. The performers, facing the same silences as the audience, have made decisions in rehearsal about what they are experiencing at that moment. They have created an "internal monologue," or an unspoken narrative that is part of the score they perform each night. These decisions are carefully choreographed and communicate the character's state of mind through posture, changes in facial expression, the direction of their gaze, and the intensity of their attitude to the other character. In short, in these short periods of time, the acting does not stop.

For the spectators, the pauses are a punctum, in the sense developed by Roland Barthes in *Camera Lucida: Reflections on Photography*. "The *punctum* points to those features of a photograph that seem to produce or convey a meaning without invoking any recognizable symbolic system. This kind of meaning is unique to the response of the individual viewer of the image" (Phùng 2013: 1). The audience gains a momentary understanding that cannot easily be put into words.

This scene between Hirst and Spooner takes place about halfway through the first act. The characters, to this point, are sparring with each other through a series of reminiscences that indicate they have known each other in the past, although what their relationship was is far from certain. Indeed, the audience cannot be sure if they knew each other in the past or if they have just met. The shape-shifting quality of the dialogue keeps the spectators from coming to any firm conclusions. Rather they are kept in a state of flux, or to use Tim Ingold's words, in "a movement of generation and dissolution in a world of becoming where things are not yet given" (Ingold 2015: 147). The silence in these "pregnant pauses" immerses the audience in the tensions of the play, inviting them to imagine the dynamics of the relationship in the passive-aggressive struggle for power. Unlike Beckett, Pinter maintains fluency through the pauses by engaging the audience in imagining the tensions between the characters. Franz Xaver Kroetz, on the other hand, sprinkles his dialogue with silences for a very different purpose.

Franz Xaver Kroetz: *Farmyard*

The *Farmyard* narrates the existence of subsistence farmers in Germany at an unspecified time and place. There are four characters: the farmer, his wife, their child Beppi, and a farmhand, Sepp. The farmhand impregnates Beppi. In retaliation the landowner shoots his dog, "an eye for an eye." Wanting to terminate the pregnancy, he turns to his wife who is ultimately unable to perform the operation.

The play ends with the beginning of Beppi's labor. Kroetz writes Beppi as if she is cognitively impaired. Nothing is said in the play to confirm this diagnosis, but she seems to be unable to make decisions for herself. The decision to portray her in this way is, I believe, made to explore the politics of property, class, and misogyny, through self-perpetuating ideologies of oppression.

Ideology is defined in the *Oxford English Dictionary* as "the set of beliefs characteristic of a social group or individual." The French philosopher Louis Althusser develops this concept in his critique of the Capitalist mode of production. He argues that the state has the function of reproducing the system (political and economic) through teaching its citizenry a set of beliefs and values that make the dominant system seem to reflect the natural order. "The reproduction of labor power requires … its submission to the rules of the ruling ideology" (Althusser 1971: 132). The work of educating the populace into accepting the dominant belief and values takes place through **Ideological State Apparatuses** (ISA) such as education, religion, and the press. Kroetz, by making Beppi a "hollow" character, foregrounds the effects of ISAs through the ways in which she is treated by the other characters. For instance, Beppi is as much the property of the farmer as the dog is the property of Sepp.

Kroetz uses silences in a very different way from Pinter. In his notes to the director, he specifies the length of time for pauses. "(*pause*) within the dialogue = caesura of roughly ten seconds/(*pause*) = silence of at least twenty seconds/(*long pause*) = silence of at least thirty seconds" (Kroetz 1976: 38). Thirty seconds of "dead air" (in radio jargon) is a long time between pieces of text. In this short scene, Sepp takes Beppi to a county fair.

> **Sepp** I'm gonna have a beer. I'm gonna go inside cause no one's serving out here.
> Coming?

Beppi (*nods*)

Sepp (*goes inside the restaurant*)

Beppi (*uncertain, looks around. Then straightens her dress, neckerchief, pulls up her knee socks, etc. Then folds her hands on the table*)

Sepp (*comes back*)

It's coming right away. (*He sits back down.*)

(*long pause*)

In five years if I'm lucky—I'm gonna retire (*pause*) then I'm a free man. No one's gonna tell me what to do, no way, nothing, no way.

(*long pause*)

Then I'm going to the city and get an apartment. Afterward you can come if you like.

(*pause*)

To the apartment, in the city. (Kroetz 1976: 47)

I encourage you to read this scene aloud, paying strict attention to Kroetz's timings. What do you imagine the characters are thinking about during the pauses? My guess is you will find there is no continuous action, no tension between the characters. In short, there is no "movement of generation and dissolution" for the spectators to engage in order to develop a deeper understanding of the characters or the action. Nothing is happening in those pauses, other than that they are sitting, and Sepp is musing about the future while waiting for the beer to be served.

The silence, in this instance, is an irritant. We look for the interior monologue found in Pinter, only to discover that there isn't one. This leads to a set of questions that are completely different from those that Spooner and Hirst inspire. Instead of interiority, there is only exteriority. Instead of looking deeper into the psyches of the characters, we are forced to the surface of the *mise en scène*. While there is no generation and dissolution between the characters, there is between the actors and the audience. The spectators follow what takes place between Sepp and Beppi. They hear and understand their words but are stymied by the lack of interiority.

As they sit in seemingly interminable silence, waiting for something to happen, the spectators confront the fact that the characters don't have compelling interior lives. There is tension between the desire to know more about the characters, who *do* engage our interest, and the empty silence. This is not to say the characters are mere puppets. They are interesting and experience emotions associated with desire, pain, and loss. But, as time passes, what captures attention is the impoverished lives of the characters and the destructive force of living in poverty. The actors can't develop an interior monologue because there is no tension or suspicions to evaluate, only counting the seconds until it is time to say the next line. This is not to say that there is no correspondence between the character/actor and the spectators. It is not, however, one of immersion into the dynamics of their interactions, but one of confronting their silence.

In both Pinter and Kroetz, silence offers an opportunity for reflection, or for the activation of what neuroscientists are calling the **default mode network**. When doing functional magnetic resonance imaging (fMRI) scans, the technicians need to first establish a baseline of brain activity. This is, to all intents and purposes, a steady state image of the brain at rest. The stimulus is induced and the resulting images map which areas of the brain "light up." They then subtract the base line information, and the remainder is the effects of the stimulations. Scientists began to wonder what takes place during this baseline working state of the brain. One "prevailing view is that the default mode or 'conscious resting state' involves retrieval and

Task 4

In groups of three, two will perform the scene from the *Farmyard*, while the third person times the pauses and silences according to Kroetz's directions. Try to fill the time developing a logical internal monologue from one statement to the next, being sure to focus only on the line of thought that will lead you from one line to the next. Perform what is said in the stage directions, especially for the one playing Beppi.

manipulation of past events, both personal and general, in an effort to solve problems and develop future plans" (Greicius et al. 2003: 257). The silence and pauses in the theatre allow the audience time to think about what they have just witnessed and how it fits in with what has happened previously and develop generalities about what takes place next. It is also in these breaks in the action that an emergent understanding of the production can begin to form.

What emerges in Pinter is very different from what arises in Kroetz. One provides character insights, the other a glimpse into the effects of poverty on life. Maria Irene Fornes uses silences to define character relationships in a different way.

Maria Irene Fornes: *Mud*

Mud is about the eternal triangle: two men vying for the same woman (Fornes 2001). It is, in many ways, similar to *Farmyard*. The three characters live in a rural area, do not have much money, and are not well educated. The writing is much more poetic, however, and the action borders on the surreal. Fornes uses pauses in significantly different ways, specifically in terms of *when* she uses them. Kroetz's characters are static through many of the silences, while in Fornes's play the characters are extremely active, except at the end of each scene when they freeze briefly before preparing for the next scene.

Mae and Lloyd have been living together for a long time. Lloyd was brought into the house by Mae's father, who decided for unknown reasons to take care of him. Mae continues to tend to his needs, sexual as well as emotional and material. The man is ill at the beginning of the play, although the cause of his illness is not known. He refuses to see a doctor. Mae is teaching herself to read in addition to taking care of the house. She brings Henry home, and they begin a relationship, Lloyd is no longer able to meet her physical needs.

The following is from the end of Act One. In the previous scene Mae has explained to Henry why she lives with Lloyd. The following takes place sometime later.

Scene Eight

Lloyd *gets up and places his chair by the table.* **Mae** *places the notebook pencil and textbook on the mantelpiece. She places the dish with string beans center and sits. She snaps the beans.* **Henry** *walks behind* **Mae** *and covers her eyes. He takes a small package from his pocket and puts it in the bowl.*

Mae What is it? (*He uncovers her eyes. She unwraps the package. It is lipstick.*) Lipstick … (**Henry** *pushes the lipstick out of the tube. He takes a mirror out of his pocket and holds it in front of her.*) A mirror. (*She holds the mirror and puts on lipstick. She puckers her lips. He kisses her.*) Oh, Henry.

(*They freeze.*)

Scene Nine

Mae *places lipstick, mirror and dish with string beans on the mantelpiece. She places textbook center and sits.* **Henry** *places the paper and lipstick cover on the mantelpiece. He takes the newspaper, turns the left chair* *toward the down-left corner and begins to read, leaning his elbow on the table.* **Lloyd** *sits on the floor, down of the right chair with his arm leaning on it.*

> **Mae** (*reading*) This is a hermit crab. He is called a hermit because he lives in empty shells that once belonged to other animals. When he is little he likes to crawl into the shells of water snails. When he grows larger he finds a larger shell. Often he tries several shells before he finds one that fits. Sometimes he wants the shell of another hermit crab and then there is a fight. Sometimes the owner is pulled out. Sometimes the owner wins and stays.
>
> (**Lloyd** *lifts himself up and looks at Henry. He mouths a curse.* **Mae** *turns to look at* **Lloyd** *and then looks at* **Henry**. **Henry** *turns back to* **Mae**, *then he looks at* **Lloyd**. *They freeze.*)
> (Fornes 2001: 27–29)

There are longer scenes with traditional dialogue and with speeches of different lengths, but these give an idea of what the play is like. The sentences are fragments, except when Mae is reading from the textbook. The short sentences, repetition of words, and simplistic information about hermit crabs indicate her reading skill is that of a beginner. The image of the hermit crab in conjunction with the stage directions after she has finished reading implies that what is happening in the house mirrors the animals' conflict when they fight for the right to possess the shell/house.

More to our purpose are the stage directions and the freezes. The former are specific and brief, suggesting there is to be minimal extraneous movement. The movement from one scene to another is not only efficient, but also done in the full view of the audience. There is no indication the lights are to change, and it seems clear that

Fornes wants the changes to be visible. There are Brechtian overtones in the decision to lay bare the workings of the theatre, to destroy the illusion created by the narrative. There are also nods to Brecht in the episodic form of the play. The text consists of short scenes, creating a disjunctive rhythm in the playing of the action, partly created by the movement from scene to scene and the economy of the dramaturgy. When there are longer speeches, the attention of the audience will be drawn to them because they are unexpected. We will address this issue in the next chapter.

The stage directions also punctuate the speeches. Unlike Pinter and Kroetz, the actors are not giving the spectators an opportunity to reflect on what is taking place, although there may be time for some of that. Their attention will be drawn to the tasks that the actors are performing. These movements and the way in which they are performed can give insights into the characters and the quality of their lives. The almost equal emphasis placed on movement and text keeps the audience from becoming too comfortable because one is never sure when one will be replaced by the other. There is a further contrast between moving and speaking: almost all the lines are spoken without movement, emphasizing the difference between the two modes of communication.

The freezes offer a third way of relaying meaning. The director has the task of creating stage pictures that intensify the image to communicate the importance of the scene just completed and the one to come. In Scene Eight, the focus is on the nascent physical relationship between Mae and Henry in the absence of Lloyd. In the next scene, all three are onstage when Mae reads the life of hermit crabs. The significance of the metaphor is not lost on the characters, who look at one another, with Lloyd's unspoken curse setting the tone for the end of the act, creating unease and curiosity in the spectator.

Task 5

When talking to friends who are not in the class, become aware of the pauses and silences in your conversation. Think about how they signal a break in the dialogue, and when there is a searching for a new topic of conversation. What does this say about how we communicate every day? Report to the class.

Conclusions

Not all texts come prepackaged on the printed page. Katie Mitchell took Virginia Woolf's *The Waves* as a pretext for her inventive, intermedia performance. Moises Kaufman and members of the Tectonic Theater Project used interviews with those surrounding the brutal murder of Matthew Shepard to create *The Laramie Project*. Herbert Blau and his Kraken company used the history of an ill-fated wagon train crossing Donner's Pass in the winter to tell the story of cannibalism as a means of survival. There are numerous other examples, some referenced here such as the work of Robert Wilson, Simon McBurney, Ariane Mnouchkine, and Romeo Castellucci who have created powerful pieces of theatre without depending on a written script. Nevertheless, the vast amount of theatre remains indebted to the creativity and artistic visions of playwrights.

We tend to think the work of a playwright as lines to be spoken by actors, and that is certainly a major part of their work. But as I hope to have shown there is equal emphasis on what is *not* said, that takes place in the pauses and unspoken stage directions that enhance and inform the telling of the story. It is in the synergy of word, silence, movement, and gesture written or suggested by playwrights that the actors, directors, dramaturgs, and designers find the tools necessary to elicit performances with the power to move audiences to laughter or tears or deep reflection. In the next chapter, we cross the immeasurable distance that separates the audience from the *mise en scène* of the performance by addressing the thorny question of aesthetics.

6

Aesthetics

Task 1

Before reading this chapter, think of an experience that you would describe as aesthetic. Answer the following: what did the experience consist of; what were the qualities of it that made you say it was aesthetic; and what was the overall effect of the event, intellectually and emotionally.

Aesthetics is traditionally defined by philosophers as an attempt to understand the unique qualities of beauty and the effects they have on the beholder. Others hold to the common claim that beauty is in the eye of the beholder. I know not everyone agrees with what I find beautiful. That doesn't mean it isn't drop dead gorgeous; just that it is *my* drop dead gorgeous. Does this mean aesthetics is totally subjective? Along with aestheticians, art historians and theorists try to convince us otherwise. There are certain works of art that are more or less recognized as standards of beauty, at least to a particular culture. Take the Sistine Chapel, for instance, or Bernini's *Apollo and Daphne* or Georgia O'Keefe's *Black Iris* as examples of works of art that are considered aesthetically pleasing. Neuroscientists are now entering the fray by trying to define the cognitive processes that provide the pleasure associated with the beautiful.

Neuroscientist V. S. Ramachandran argues that beauty is not to be found in the work of art, but that the work has inherent qualities that in their effect on the beholder give rise to aesthetic pleasure. This is what Ramachandran claims in his discussion of **neuroaesthetics**. He outlines nine laws "to explain why artists create art and why people enjoy viewing it" (Ramachandran 2011: 241). They are: contrast, isolation, perceptual problem solving, "abhorrence of coincidences," orderliness, symmetry, metaphor, grouping, and peak shift. The evidence he brings to bear is largely based on neuro-imaging, such as fMRI or Pet scans, and other scientific findings, such as the physiognomy of seagulls. When employed judiciously by an artist, obeying these laws will light up the pleasure centers in the brain, giving rise to an aesthetic experience. "When self-amplifying echoes between these layers of visual processing reach a critical volume, they get delivered as a final, kick-ass 'Aha!'" (Ramachandran 2011: 244). There are several problems with this equation. The argument is based on a limited range of visual arts. The examples he gives in defense of his thesis are drawn by and large from Indian art, which is understandable given his ethnic background. Less excusable is that many of his images are of curvaceous women, whose poses accentuate their eroticism and exoticism. Moreover, they are often taken out of context. He discusses two drawings of horses, one by a girl with autism, and the other by Leonardo Da Vinci (222). He finds the young woman's depiction more aesthetically pleasing and therefore a better work of art. Ramachandran doesn't consider why Da Vinci made the drawing or what the standards for beauty were for that time. He is only interested in defining the universals of aesthetic appeal in contemporary culture. Most disturbing, however, is the claim that if a work of art doesn't fit his definitions, it is not art. "White paint on a white canvas could hardly be called art" (219). He places himself, a scientist, as an arbiter of what is and is not considered art, despite the work of Russian Formalist Kazimir Malevich, Abstract Expressionist Victoria Kloch, or the ready-mades of Marcel Duchamp. I am not questioning the validity of his categories or their potential usefulness in understanding art. Nor am I questioning the science. What I do assert is that they are not necessary or sufficient for articulating a theory of the aesthetic.

Vittorio Gallese is one of the scientists who discovered mirror neurons in macaque monkeys. Mirror neurons, if you recall the earlier discussion, fire when the animals execute a task as well as when they see it performed by another, although at different intensities. In collaboration with art historian David Freedberg, Gallese looks at emotion and empathy in the experience of art (Gallese and Freedberg 2007). More circumspect in their claims, they argue that viewers of visual art are affected emotionally and empathically by certain aspects of art. One of their examples is the work of Lucio Fontana, known for his cut paintings, in which he slashes the canvas with a knife. Gallese and Freedberg argue that when we see the work of art, the mirror neurons are activated as if the viewer was cutting the canvas. This imagined act, according to the authors, initiates emotional and empathic responses in the viewer that define the aesthetic experience. The authors recognize that this is not true for all works, and unlike Ramachandran, do not put limits on what qualifies as art. However, there are other aspects of Fontana's work, such as the overall composition, the effect of lighting, and the context in which it is displayed, that they do not consider. Each of these affects the experience and may play just as large a role as the mirroring mechanism's response to the residue of the stroke of the knife.

The unspoken, common denominator for both of these theories is that seeing art is not only an embodied experience but also one that is embedded in everyday reality and takes place over time. In what follows, we will look at the phenomenological experience of viewing art, whether or not we consider it beautiful. In fact, I intend to avoid the question of beauty because of the subjective nature of our responses to different works of art. Moreover, works by Mozart, Beethoven, Debussy, Shakespeare, Moliere, or Sophocles are works of art, but involve a different set of tools for analysis. They constitute art, as do dance and other forms of performance, and therefore must be considered in any theory of aesthetics. Instead, I will adopt Anoka Faruqee's reading of Kant's understanding of the aesthetic experience: "aesthetic judgment is an experience where the imagination presents an image to the understanding, which stymies the impulse of the understanding to produce an abstract concept that will effectively contain or render intelligible the sense particulars" (Faruqee 2009). For me, the aesthetic experience is temporal and in so being it opposes an easy determination of meaning. It is defined as much by the duration and the ambiguities arising from the event as the inherent qualities of the object or the neurological underpinnings of the mirroring mechanism. The discussion will begin with a case study.

The Concept of the Face: Regarding the Son of God

I saw Romeo Castellucci's *The Concept of the Face: Regarding the Son of God* performed by the Societas Raffaello Sanzio in February 2013. The performance is divided into three parts, only the first using spoken text. The setting, which is in full view on entering the theatre, is divided into three areas. Stage right is a living room, center stage a kitchen table and chair, and stage left a bed and side table. All are parallel to the front of the stage and are in white, chrome, and light wood, perhaps birch. The back of the stage consists of a floor to flies detail of the *Salvator Mundi* by Antonello da Messina depicting the Ecce Homo, or the passion of Christ after being scourged. Despite the suffering he endured, there is no sense of pain, condemnation, or forgiveness—the face is simply passive.

The first scene depicts a middle-aged son taking care of his incontinent father. They move from stage right to stage left, and in each of the three areas of the stage, the father soils himself, each time more violently and graphicly than the last. The first is a brown stain on the back of his white robe and the last what can best be described as projectile diarrhea, which was accompanied by the aroma of feces pumped into the theatre. Each time the son washes his father with loving compassion. The father is increasingly humiliated by his inability to control his bowels, while the son, despite his caring, grows increasingly frustrated by the repeated need to clean his father. Towering over these actions is the passive Ecce Homo. The first section ends with the father pouring disinfectant over himself and sitting on the bed.

The realistic representation of feces was so repulsive that I found it impossible to look. My gaze shifted from the action to the image of the Christ. It lingered there briefly until the disgust receded and I was able to watch the action on stage once again. It gradually became clear that Castellucci was creating a triangle between the two men and the painting. The effect of the disgust played against my feelings of empathy, but the looking away from the men to look at the painting changed my experience of the image. No longer admiring the work of art, I was struck by the indifference of the face as it looked down at the struggles of the two men. The juxtaposition of image and characters emerged as a critique of the Catholic Church. The Christ, who is the epitome of compassion and forgiveness, has become an impassive observer. The power of the church, implicit in the scale of the painting, has lost touch with the suffering of human beings.

The second part of the performance was shorter, probably ten minutes, but much more violent. Seven teens came onstage one after the other, opened their backpacks and took out hand grenades that were hurled at the image of the Christ. There was no regularity in their actions; each threw at their own pace. As each grenade hit the face, the "explosion" was amplified and reverberated. As more were thrown, the level of noise increased and became cacophonous. The act itself seemed senseless, because there was no overt anger in the throwing, in fact no reason at all. It was just a task to be completed. It was mesmerizing because of the lack of anger in the young people and the aural violence of the action. What became meaningful was the lack of impact of the grenades hitting the face. There were no dents, bruises, or cuts, just the continued impassivity of the gaze. After they finished, they walked off, except the first one to enter. He paused before leaving and looked at the old man on the bed covered with disinfectant. Another triangle was created: the youth on one side of the stage, advanced old age on the other, and the omnipresent image of Christ between them. The silence after the cacophony and the stillness of the two performers gave time for reflection on the relationship of religion to human beings throughout their lifespan.

The final act of the performance was the destruction of the image. In a scene that had its own beauty, the painting began to dissolve as a liquid appeared to be poured over it. The colors faded and the fabric disintegrated. Performers rappelled from the flies, ripping the fabric, revealing at first the words "The Lord is my shepherd," and then "The Lord is not my shepherd."

The Concept of the Face was a powerful experience, and is one to which I frequently return in my work. It started me asking: what is an aesthetic experience that can contain such repulsive images and violence? Because for me there is no question: it was an aesthetic experience. In the next section, this train of thought will be discussed in greater depth, beginning with the theories of Bertolt Brecht and Antonin Artaud.

Defamiliarization and cruelty

The Russian Formalist Viktor Shklovsky, who coined the word *ostranenie* or **defamiliarization** in English, sought to understand what makes art brilliant. His great insight is that, in art, ordinary things are rendered strange.

> The technique of art is to make objects 'unfamiliar', to make forms difficult, to increase the difficulty and length of perception because the process of perception is an aesthetic end in itself and must be prolonged. *Art is a way of experiencing the artfulness of an object: the object is not important.* (Shklovsky 1988: 20)

When Duchamp turned the urinal on its end, it ceased to be the functional object that we know. It required time to perceive what it was and then ask what it was doing in an art gallery. What was mundane stimulated curiosity, and in this case outrage at the defamation of what was traditionally considered fine art.

Brecht was influenced by Shklovsky in the development of this theory of **Verfremdungseffekt**, normally translated as alienation effect. I prefer the Russian's more positive formulation: defamiliarization. In

---

Task 2

Look at works of art by Rene Magritte, Salvador Dali, and/or M. C. Escher. Discuss how these artists have taken everyday objects and rendered them unfamiliar.

his essay "The Literarization of the Theatre," Brecht writes that the theatre needs to involve an "exercise in complex seeing" (Brecht 1964: 44). Marcel Proust provides an example. He takes an unexceptional cookie, the petite madeleine, and from the taste of it dipped in tea elicits memories that unfold in *Á la recherche du temps perdu*. "Undoubtedly what is thus palpitating in the depths of my being must be the image, the visual memory which, being linked to that taste, is trying to follow it into my conscious mind" (Proust 1981, 1: 49). The simple act of dipping a cookie into a cup of tea changed the narrator's understanding of an everyday event because it opened onto a raft of memories about his life. The same thing happens, although in a different register, to the reader as this singularly unexceptional event opens onto the memories that make up the extended reverie that is the novel.

Complexity in seeing is not simply a switch that can be turned on. It requires a disfluency (see Chapter 5), a break in the flow of experience, and, at least for Brecht, a reorientation of perspective. "It is perhaps more important to be able to think above the stream than to think in the stream" (Brecht 1964: 44). The disruption breaks our immediate connection with the object or event, creating a metaphorical distance between the work and the spectator, moving the viewer from active engagement to contemplation. American theorist and director Herbert Blau pointed out during an NEH seminar in 1987 that "to think above the stream," you first need to be in the stream. That is, spectators need to be immersed in the reality of the moment (the familiar) and only then can it be made unfamiliar in contrast to its former state.

When the old man entered in *The Concept of the Face*, there was no indication of his incontinence. He was simply an old man coming into the living room. It was only when he turned upstage that the stain became visible, shifting considerably our understanding of not only the old man but also the production as a whole. Similarly, the painting upstage was a beautiful object to look at, and only became significant when the incontinence forced me to look at it. A dialectic was then initiated between the pathos of what was taking place on stage and the passivity of the face. In being pulled out of the "stream," the perspective on the action changed, allowing for a moment of reflection and the gradual emergence of meaning.

The disruption of expectations in *The Concept of the Face* cannot only be attributed to the defamiliarization of the everyday, because there was a violence in the moments of incontinence that were cruel in the sense of Artaud's theatre. Artaud sought to breach the boundaries between the realm of the repressed and everyday life. Perhaps, the purest of Surrealists, he felt intimately the force of civilization, including the structures of consciousness, to resist the expression of impulses and instincts that we have grown to call the irrational. Artaud and the Surrealists felt that society could not be healed until these repressed energies were released and allowed expression in our lives. To do this, he felt that we needed to break the hold of language on our theatres and present images that challenged our deepest held beliefs. His belief in the stranglehold of language came from his discovery that, in writing poetry, he was unable to communicate his innermost passions. He felt that language "stole his breath away." It is through sonorizations and violent images that there is the potential to break through the passivity of the spectator and a true theatre be realized, one that releases theatre's double, the always present energy that haunts the surface of the text. The disturbing images of the defecations, the cacophony that accompanied the throwing of hand grenades, and the shredding of the image of Christ, while not reaching the magnitude of Artaud's vision, were close approximations of that drive to contact the spectator in a nonverbal, visceral way that is cruel and liberating.

The examples of Brecht and Artaud are not to be found in every production. I'll leave it to you whether or not this is a good thing. But they define two points that, with the tradition established by Stanislavki, create a field for identifying the tensions that exist within most performances, particularly in the post-dramatic. This formulation does not diminish the power of what the spectators experience, but it provides a tool for understanding the nature of the aesthetic experience, whether it is art (Egon Schiele), music (Nine Inch Nails), dance (Degenerate Art Ensemble), performance art (Hermann Nitsch), or theatre. However, defamiliarization, cruelty, and realism do not in themselves explain the aesthetic experience on either a personal or cultural level.

The horizon of expectations

The defamiliarization, cruelty, and the realist tradition depend on an antecedent, such as, in *The Concept of the Face,* the establishment of the old man as a character without an infirmity. With his appearance a *horizon of expectations* is established based on prior experiences with old men. I am extrapolating the concept of a horizon from the reception theory of Hans Robert Jauss, a German literary critic. For Jauss, every age approaches a text based on their experiences of reading similar kinds of works and the ideological values inherent in the culture. "The new text evokes for the reader (listener) the horizon of expectations and 'rule of the game' familiar to him from earlier texts, which as such can then be varied, extended, corrected, but also transformed, crossed out, or simply reproduced" (Jauss 1982: 88). It is not only experiences with other texts that define the horizon of expectations, since it is "received and judged against the background of other works of art as well as against the background of the everyday experience of life" (41). There are then multiple horizons formed by experience with theatre and art in general. But it is also mediated by our lived experience, which, as we saw in Chapter 2, is determined in part by the beliefs and values we adopted as we grew up.

I went to see *The Concept of the Face* with a history of going to the theatre, which included an understanding of proper etiquette, knowledge of different acting and directing styles, and the traditional arc of action. I also had knowledge of the company based on work seen previously and some research for courses I taught, which included YouTube videos. This understanding was interwoven with an interest in more experimental forms of theatre as a director and audience member. These particularities determined, to a large extent, my state of mind when the performance began. This fluency, fueled by a set of expectations, was broken by the appearance of a brown stain on a white bathrobe.

Each incidence of incontinence was a disfluency that broke with my expectations. If each event had been identical, the effect would have diminished. Increasing the amount of feces intensified the discomfort, making even what was becoming familiar through repetition unfamiliar. The final break in continuity was the father unexpectedly pouring the disinfectant over himself. The final two sections also held surprises that undermined my expectations. The teens throwing objects at the painting was a surprise initially, but quickly became a pattern, although the duration of the action and cacophony increased my discomfort as a spectator. What was unexpected was the disintegration of the painting and the emergence of the words at the end. It was not only unexpected but also it had a haunting beauty as the solidity of the image fell in shreds, revealing the final words.

Every theatre experience is going to engage with the horizon of expectations. Every performance provides what we can call the *horizon of the text*, or the structures that give rise to literal and affective meaning as well as the ideological values and beliefs of the author(s) and production company. Audience members' experience in the theatre forms their *horizon of experience*. This coming together of the various horizons will either fail to meet, match, or exceed what is anticipated. The event is going to reinforce or modify, marginally or exponentially, what is expected from going to the theatre. The horizon is ever changing. Moreover, it is not monolithic but made of subsets of expectations about texts, performances, design elements, and the theatre space itself. The beginning of Mallory Catlett's *This Was the End* (see Chapter 4) began with the pulsing of the set. By reducing and expanding a video projection of the wall on the wall, the wall appeared to breathe. I had never experienced such an effect. It not only disrupted my expectations of what was to come but also forever changed what I can expect from the integration of technology and scene design.

We go to the theatre to have our expectations exceeded. If the production fails to meet what we define as the best that theatre can offer, we come away disappointed. If there is a match between them, we are bored because we wanted something more, something different. It is only when there is a surprise, when there is an innovative patterning that brings the audience into the stream of the performance and pulls them out of it through disfluencies that give rise to reflection, that

---

Task 3

Share with a partner what you hope to experience when you go to see a performance, preferably the theatre but other forms of art as well. Remember what it was about a specific event that failed to meet, met, or exceeded your expectations. Be specific in describing the experience. What do you feel when you recall the experience?

we experience the pleasures of the theatre, be they happy, sad, or any other kind of emotional and intellectual response.

Temporality and aesthetics

Performances are temporal art forms. Whether as brief as Beckett's *Breath* or John Cage's *4'33"* or as long as the seven-day ritual performances of Herman Nitsch's Orgien Mysterien Theater, time is integral. Time was discussed in Chapter 4 as it pertains to theatre, and again in Chapter 5 as part of the performance of text. Here, we develop those ideas as part of the aesthetic experience with specific reference to expectations, fluency, defamiliarization, and disfluency. Before turning to theatre, however, an experience with visual art may be useful. The reason for taking this detour is the relative immutability of sculpture and painting in comparison to theatre and performance.

Attending a conference in Dallas, Texas, a number of us decided to visit the Dallas Museum of Art. As we walked down the main corridor of the building, my eye was drawn to a bright orange lozenge that seemed to explode off the wall. I left my colleagues and walked into the space. There was an antechamber before entering the main gallery and two paintings on opposite walls, one by Jackson Pollock and the other by Lee Krasner, two major Abstract Expressionists who happened to be married. The main gallery, continuing the theme, consisted of a range of painters who worked in that era. There were paintings by Franz Kline, Jasper Johns, Bruce Nauman, Willem De Koonig, and others. The painting that caught my eye was by Mark Rothko, an artist I have long admired. All the works were paintings hung vertically on the wall. I looked at them all, some more intently than others. Each one required a different way of looking at it depending on the emphasis given in the composition to line, color, and texture. Only later did I realize that the shift in approach required by each painting was creating what I now call a visual noise. There were so many stimuli that I was bubbling with information and perceptual impressions.

Completing that room, I turned into the next gallery. The experience could not have been more different. William Smithson's *Mirrors and Shelly Sand* is a horizontal rectangular installation consisting of fifty 12-by-48-inch panes of glass supported by a mixture of shells and sand (Smithson 1979–80). What was most striking, however, was the relative silence of the room. Only the hum of the HVAC kept it from being totally quiet. The lack of noise was reinforced by the horizontality of the work of art, which was more peaceful to the eye than the verticality of the other works. Placed in the center of the room, the viewer was encouraged to walk around the piece, exploring how the light played with the transparency of the glass. Patterns shifted; shadows lengthened and shortened. At one point, it came to mind that they are both made up of the same material: silica. It was only after I slowly worked my way around the piece that I stopped to read the sign describing it and realized that Shelly's poem *Ozymandias* was the inspiration for Smithson's piece.

I am recounting this experience in detail to underscore the stages of moving through the rooms of the gallery and, by extension, the temporal aspect of the event. The orange lozenge of Rothko's work changed my focus of attention, drawing me to the piece. Each painting required at least a small amount of time viewing it, only to decide I didn't want to look at it, and each established a set of expectations for the next, which were disrupted by the demands of the new work. Certain categories of seeing were common to all, but the specificity of each work added new information that altered, however slightly, the horizon of my expectations. The mass of perceptual information created the visual cacophony that, surprisingly perhaps, established an erratic fluency. Disfluency asserted itself as I entered the room where the Smithson was displayed.

The sudden change in the perceptual experience drove all the bits of information held in short-term memory out of the focus of attention except for the immediate experience of the quiet and the space. This jolt put me into a state of reflection and contemplation that made all my previous thoughts irrelevant. My mind became quiet as I looked without preconception at the installation, giving me the mental space to observe the play of light on the panes of glass without conflicting information. Through observation and contemplation, I was able to come to an understanding of the experience, if not the work, that remains with me.

The sequence with which I started the section—expectation, fluency, defamiliarization, cruelty, disfluency—defines stages in the aesthetic event. All the particularities that accrued as I looked at the paintings and the generalities of what were to come when I entered the room with the Smithson were disrupted when I was met with quiet and the horizontal orientation. This experience is similar to seeing *The Concept of the Face*, where my involvement in the exchange between father and son was broken with the incontinence of the older man. Each experience broke the flow of expectations, giving me time to reflect on what was happening and to make initial steps toward an understanding. In Brecht's terms, being pulled out of the stream

---

Task 4

In a small group or by yourself think about a performance that you remember vividly, it need not be theatre. What do you remember about the event. Why do you think you remember it? Given the definition of an aesthetic experience that I have outlined here, would you call it an aesthetic experience?

in which I had been traveling led to a complex way of seeing. This, I believe, is the basis of an aesthetic experience.

Aesthetics, time, and performance

Deleuze's concept of time replaces past, present, and future with particularities, living present, and generalities (see Chapter 4). The aesthetic experience as defined here occurs when, in the living present, the particularities are discombobulated due to new perceptual inputs and the generalities of what is going to happen no longer make sense. This undoubtedly sounds cold and formulaic, lacking the emotional and reasoning response to performance. However, it offers a more concrete foundation than the abstract and subjective idea of beauty. It also offers a structure for evaluating the experience. Perhaps some examples are in order.

My wife and I saw *Other Desert Cities* by Jon Robin Baitz on Broadway. It featured a wonderful cast, including Stacy Keach and Stockard Channing. The play is about a family at Christmas. The brief synopsis provided by Dramatist's Play Service runs:

> Brooke Wyeth returns home to Palm Springs after a six-year absence to celebrate Christmas with her parents, her brother, and her aunt. Brooke announces that she is about to publish a memoir dredging up a pivotal and tragic event in the family's history—a wound they don't want reopened. In effect, she draws a line in the sand and dares them all to cross it. (Baitz 2011)

At the end of the first act, we looked at each other and said, more or less, "Wow! I don't have any idea what is going to happen." There was a palpable sense of excitement because we were unable to map a set of generalities that gave us a clear path to the end of the play. We had high expectations for the second act. Unfortunately, when the play was over, all the loose ends had been tied up into a tidy resolution, leaving us with nothing to contemplate on the trip home. It failed to live up to our expectations by returning to the conventions of the well-wrought play. While enjoyable, it was not what I define as an aesthetic experience.

John Moran's *Book of the Dead (2nd Avenue)* was performed at New York's Public Theatre in 2000. It, like *The Concept of the Face*, is in three parts, but with a very different trajectory. The second section, which I will focus on, had tableaus on either side, one involving a meditation on the *Egyptian Book of the Dead* and the last part the *Tibetan Book of the Dead*. As these "bookends" suggest, the piece was about death and the possibility of an afterlife, either as a mummy or through reincarnation. They were beautifully staged using Egyptian and Buddhist iconography, with a narration by Uma Thurman.

The middle section consisted of three episodes, one set in a MacDonald's, another in a bar, and the third a homeless person living on 2nd Avenue. The intention, as far as one can know, was to recreate, for a short period of time, the experience of living in each environment. I have written about the performance in *Toward a General Theory of Acting: Cognitive Science and Performance*.

> Two actors are alone on stage, one portraying the manager of a MacDonald's franchise giving orders to Jamal, an unseen employee, and the other a waitress taking an order from an unseen customer. The speeches and movements are repeated precisely, again and again. As the sequences—speech and movement— recur, another actor enters, filling in the dialogue as he orders his breakfast. Gradually more performers enter, providing additional layers of interaction and the phrases heard at the opening of the

scene blend into and gradually become lost in the cacophony of a realistic depiction of breakfast at MacDonald's.[1] The actors are not speaking the lines, but lip-synching recorded dialogue while repeating movement phrases with absolute precision. The cyclic repetition of movements and conversations, in this and other scenes that form the second act, create a dynamic image out of the minutiae of everyday life. (Lutterbie 2011: 224)

By the time the scene ended, the image on stage was a specific and hyper-realistic reproduction of typical events in the fast food restaurant. The initial lines soon became lost in the fugue-like structure of text as layer upon layer was added. If one were to time the real-life situation, each scene would have lasted at most a minute or two. But by adding new information after every two or three repetitions, the scene with all its accretions lasted about fifteen minutes. It was a masterful dissection of breakfast time in MacDonald's and a stunning composition, musically as much as theatrically.

Walking out on to Lafayette Avenue after the performance, a small cluster of friends walking by burst into laughter. The musicality of their combined voices took me immediately back to the performance, underscoring the musicality in everyday language and the genius of Moran's work. Later thinking about the relationship between the three sections, I began to think about the poverty of the lives we live: the mundane quality of each location, and the equally mundane interaction between the characters appeared to take on a life of its own, an afterlife if you will, when the polyphonic qualities of everyday language were foregrounded. The boring activities of life were

---

[1]John Moran, *Book of the Dead (2nd Avenue)*, New York Public Library for the Performing Arts. Videotaped in Martinson Hall, Public Theatre, by Theatre on Film and Tape Archive at the New York Public Library for the Performing Arts. December 14, 2000. The Joseph Papp Public Theater/New York Shakespeare Festival, George C. Wolfe, producer.

enlivened by the energy infusing the dialogue. At the same time, there was the question of how prepared we are to die, whether there be an afterlife or not.

This was, for me, an aesthetic experience. I went in with few preconceptions. Moran was unknown to me, but from the publicity I encountered it was clearly not going to be a traditional play. The theatre was equally unknown. The particularities with which I began were determined with my entering the space and viewing the scenography that consisted of iconography of Egypt, specifically pyramids and hieroglyphics. There was very little information with which to develop generalities. Uma Thurman's narration prepared me for hearing recorded voices, and the serenity of her performance created a contemplative atmosphere that was disrupted when the second section began. One set of expectations was replaced by another, which suggested a pattern that was then replaced by another. The piece ended with a sense of wholeness, but without a clear resolution. The unexpected return to the performance with the trill of laughter added yet another dimension to the experience.

As I write about it for the second time, the first being in my earlier book, I rediscover the production, returning to past conclusions but finding nuances I had not considered before. In a sense, the aesthetic experience of Moran's work continues. It hasn't ended but returns, consolidating previous memories while adding new resonances that, like the text, add new strata of complexity. Virtually all the performances I have discussed in this book qualify for me as aesthetic experiences according to this definition. They do not die but return, sometimes intentionally and at others unbidden. They continue to stir me and bring me pleasures and pains.

Aesthetics is not about beauty, although that can certainly play a part. It occurs when time is taken to interact with a work of art, paying attention to what it has to offer and the response it evokes in you. It is an immersion for which there are no words, at least not immediately. They may come after. Instead, there is something you did not, could not have expected and absorbing what is perceivable is all you can do, that is, until your attention is drawn elsewhere. That new thing that now captures you continues to take you along, providing new perceptions that enrich the experience by creating a fluency or disfluency. It is only at some later time that you can begin to piece things together and develop an understanding of the encounter. It is in the act of contemplation and reflection that emotions return, as do thoughts, and it becomes possible to put part of it into words and meaning emerges.

By exceeding expectations, the experience makes a lasting impression that you cannot easily shake. It returns because there is no easy way to categorize the effect it has on you, and there is no immediate way of filing it away with other encounters with art. You must sit with it, interrogate it, and allow it to return to mind. In that moment of recognition, you realize it was over too soon, and you want to live with it again, if only in memory.

At the end of Richard Maxwell's *The Evening*, the set that you have been looking at throughout the piece disappears and fog starts to fill the stage. The woman who is leaving abusive relationships begins to walk from down stage left to upstage right. She is wearing a coat made of light-colored feathers. By the time she gets to her exit, she is no longer visible, enfolded in the billowing fog. An apt metaphor for an aesthetic experience.

Epilogue

Theatre is an amazing art form. When it works it has the power to move us like a Beethoven symphony or a Radiohead concert. It requires as much technique and control as a Bach fugue or a Simone Biles gymnastic performance. It can mirror reality or present a stream of images that refuse narrative. No single book can begin to encompass the magnificent vastness of theatre and performance. This one is no exception. Whole realms of theatrical practice around the world are not represented here. The theatre of Africa, Asia, the Middle East, South America, and Australia hardly figure, if at all. Repressed voices are not raised here: immigrants, refugees, women, LGBTQ+, and those discriminated against because of race or ethnicity. It is not because they are not as significant as those represented here, in fact quite the opposite. There is shame and regret in this, stemming from the limitations of my knowledge. Without wanting to sound glib: so many books, so little time, and my clock is running down. As the Replicant says in Ridley Scott's *Blade Runner*: "I've seen things you people wouldn't believe. Attack ships on fire off the shoulder of Orion. I watched C-beams glitter in the dark near the Tannhäuser Gate. All those moments will be lost in time, like tears in rain. Time to die" (Parsons 2013). While I hope not to die in the very near future, there is more to be said, especially about things I do not know.

Theatre, wherever it is performed, requires people. Even Beckett's *Breath*, which requires no performer, still needs someone to set the stage, turn on the lights, run the sound, and open the house. Every one of them brings to the work a perspective that has been developed over a lifetime of engaging with others and, in the process, developing values and beliefs that are central to their standards for work well done. Much of this is determined by the (sub)culture in which they grew up. In the nature/nurture debate, the nod goes to nurture.

There is much that is hardwired in the brain, that magnificent organ that is both structured and open to change. But it is also true that our memories, the foundations of values and beliefs, arise from our interactions with the world. Those pathways through the neural networks are plastic. They can give form; they can receive form (Malabou 2008: 5). The more often they are accessed the more difficult they are to change, but change is always possible, if there is a will.

It is true that as we age we become more set in our ways, but we can also develop a mindset that seeks out new experiences and new ways of thinking that embraces change. The arts provide an arena, for practitioners and spectators alike, to engage with creative expression. It takes determination and willingness to be disappointed but the payback is significant. I frequently teach a Theatre in New York course. It is a little hyperbolic because we don't go to Broadway and only occasionally dip into Off-Broadway. Part of the reason is cost of tickets, but the more important objective is to introduce students to experimental works, most being created by their peers and emerging artists. They are all wary to begin with, but by the end of the term, they value the experience of having their horizons expanded. They seldom like the same things, but they now know what they like and don't like.

This cannot be done through books, although they help. It requires putting your body in the seat and paying attention to what is going on.

I tell students tempted to turn on their phones that if they are bored, they should resist the temptation and think about why it is boring. These are the first steps in developing a personal aesthetics and, perhaps more importantly in gaining confidence in what you value and what you do not. There is no special technique required. It is how our cognitive processes work, all day, every day. We are embodied: we act on the world as it acts on us. We extend ourselves into the world in which we are embedded. Maurice Merleau-Ponty defines the flesh as the extent of our senses, and not the limits of our skin (Merleau-Ponty 1968: 135). Our flesh reaches as far as we can hear, smell, and see. In so doing, we intertwine with the flesh of others, and that is what makes acting and spectating exciting and rewarding: taking the time in a space to engage with others without fear of reprisal. We can experience who we are and allow ourselves to be affected by the event.

That is the value of cognitive science to the arts: to get a better understanding of who we are and the world in which we live. It provides us with tools to understand how we learn and develop expectations and explore what happens when those expectations go unmet or are exceeded. But it is not a one-way street. Science can learn from the arts, as much if not more than we can gain from reading science. Here are two examples.

"Imagining Autism" was founded by Nicola Shaughnessy and Melissa Trimingham at the University of Kent in the UK. They both have children with autism. Feeling that children on the autism spectrum were not being given opportunities to develop their creative and social skills, they developed a program that goes against the normal protocols of working with autists. The standards call for a controlled environment with few distractions and a regime of repeated activities. Their experiences with their children suggested that there is a better approach to interacting with autistic children. Instead of limiting stimuli, they believed that short sessions of intense stimulation could help address some of the symptoms of autism. Pods were developed each with a different environment: the Arctic, the ocean, the forest, and the city. There were props, puppets, digital projections, and costumed caregivers, trained to respond should the students have an episode. Each intervention lasted approximately one hour and was driven by what the students felt like doing. They were free to explore the environment. The initial results from a study on the project were positive, with parents and teachers noticing significant differences in behavior (Beadle-Brown et al. 2018). One child began to explore his voice and language in ways he never had before. Students who did not interact well with others were sharing and joining in activities together. Empirical data is difficult to come by given the number of variables, but they are exploring new technologies for gathering data. It has the potential for altering interventions with autistic children.

Italian neuroscientists joined theatre practitioners in working with patients with Parkinson's disease over a three-year period (Mudugno et al. 2010). In addition to normative protocols for working with those with the neurodegenerative disease, subjects would go through physical and vocal exercises and short interactive scenes. Over time, the patients demonstrated greater vocal and physical control. They re-learned socialization skills and the ability to express emotion. In some instances, patients became able to lead exercises. Neither the Parkinson's nor the autism interventions claim to provide a cure, but they do show that theatre can provide relief from some symptoms and improve quality of life. It is also important to note that these studies demonstrate how, in the right circumstances, theatre can open new avenues for scientific research.

This landscape of theatre from the perspective of cognitive science is far from complete. It is a sketch, a pencil drawing, that gives a mere hint of what there is to be known. The study of cognition can open vistas on the study of theatre, leading to insights unavailable from other practical, historical, or theoretical approaches. Furthermore, it can help us understand how we interact with the world in our everyday lives and give us the opportunity to improve our lives and the world around us. It can make us better practitioners, whether we be historians, critics, or theorists. It can improve our work in the studio, on stage, and in the classroom. While reading scientific articles can be daunting, it can also help us to better understand how art works, personally, culturally, and politically.

The brain is a magnificent organ, one that we are only beginning to understand. It is capable of solving the most complex problems and performing the simplest of actions. In comparison with the brain, a robot can beat a chess master, but it can't keep from running into a wall. Science is only beginning to understand the dynamics that go into picking up a feather, a glass of water, or a ten-pound weight. Those are abilities that are beyond scientific understanding. New discoveries are being made every day, however, that increase our knowledge of how we live in the world.

Were there a different author, this book would have told a different story, focusing on different productions and different areas of scientific inquiry. Further Reading provides a selection of works outside of those cited here for those interested in further reading. It is my hope that you will seek to expand the limitations of this monograph. There is a rich field of resources that can help us all gain a better understanding of theatre and performance.

There is a brave new world at the intersection of art and science that can help us understand the creative act, making us better artists and audiences. This book does not supersede those on acting, directing, playwriting, design, technical production, history, criticism, or theory. Each of these fields adds to our understanding of theatre and performance. Science opens new horizons and new landscapes for gaining knowledge about the field of theatre studies.

References

Introduction

Etchells, Tim (2008). *Forced Entertainment: On Making Performance*. Center for Performance Research. Viewed March 5, 2018. https://www.youtube.com/watch?v=m2fRfN5U7GA

Gallagher, Shaun (2005). *How the Body Shapes the Mind*. Oxford: The Clarendon Press.

Chapter 1

Anonymous (2010). "Little West 12th Night." *Time Out New York*. Viewed June 22, 2018. http://www.timeout.com/newyork/things-to-do/little-west-12th-night

Barthes, Roland (1977). *Image/Music/Text*. Trans. and Ed. Stephen Heath. New York: Hill and Wang

Chas, Moira, and Anthony Phillips (2009). "Self-intersection numbers of curves on the punctured torus." Viewed April 4, 2019. arXiv:0901.2974 [math.GT].

Crease, Robert P., and John Lutterbie (2006). "Technique," in *Staging Philosophy: Intersections of Theater, Performance and Philosophy*. Ann Arbor: University of Michigan Press.

Damasio, Antonio (2000). *The Feeling of What Happens: Body and Emotion in the Making of Consciousness*. Portsmouth, NH: Heinemann.

Derrida, Jacques (2005). *Rogues: Two Essays on Reason*. Trans. Pascale-Anne Brault and Michael Naas. Stanford, CA: Stanford University Press.

Gibson, James J. (1986). *The Ecological Approach to Visual Perception*. New York: Psychology Press.

Goldin-Meadow, Susan (2009). "How Gesture Promotes Learning throughout Childhood." *Childhood Development Perspectives*, 3 (2), 106–111.

Heidegger, Martin (1996). *Being and Time: A Translation of* Sein und Zeit. Trans. Joan Stambaugh. Albany: State University of New York Press.

Hutchins, Edwin (2007). "Enaction, Imagination, and Insight," in *Enaction: Toward a New Paradigm for Cognitive Science*. Cambridge, MA, and London: The MIT Press.

Ingold, Tim (2007). *Lines a Brief History*. London and New York: Routledge.

Jauss, Hans Robert (1982). *Toward an Aesthetic of Reception*. Trans. Timothy Bahti. Minneapolis: University of Minnesota Press.

Kelso, J. A. Scott (2008). "An Essay of Understanding the Mind." *Ecological Psychology*, 20, 2.

Kuppers, Petra (2015 [2014]). "Swimming with the Salamander: A Community Eco-Performance Project." *Performing Ethos*, 5 (1+2), 119–35.

Lahr, John (2009). "Panic Attack: *Waiting for Godot* Back on Broadway." *The New Yorker*, May 18. Viewed June 22, 2018. http://www.newyorker.com/magazine/2009/05/18/panic-attack

Lakoff, George, and Mark Johnson (1980). *Metaphors We Live By*. Chicago: University of Chicago Press.

Lakoff, George, and Mark Johnson (1999). *Philosophy in the Flesh: The Embodied Mind and Its Challenge to Western Thought*. New York: Basic Books.

Malabou, Catherine (2008). *What Should We Do with Our Brain?* New York: Fordham University Press.

McNeill, David (2005). *Gesture and Thought*. Chicago and London: University of Chicago Press.

Merleau-Ponty, Maurice (1968). "The Intertwining and the Chiasm," in *The Visible and the Invisible*. Ed. Claude LeFort. Trans. Alphonso Lingis. Evanston: Northwestern University Press.

New, Sophia, and Daniel Belasco Rogers (2010). "Me, You, and Everywhere We Go: Plan B." *Performance Research*, 15 (4), 23–31.

Peirce, C. S. (1998). *The Essential Peirce, Volume 2*. Peirce edition Project. Bloomington: Indiana University Press.

Saussure, Ferdinand de (1998). *Course in General Linguistics*. Trans. Wade Baskin. Peru, IL: Open Court Publishing Company.

Schechner, Richard (1977). *Essays on Performance Theory 1970–1976*. New York: Drama Book Specialists.

Taubes, Gary (1990). "The Cold Fusion Conundrum at Texas A&M." *Science*, 248 (4961), 1299–304.

Chapter 2

Althusser, Louis (1971). *Lenin and Philosophy and Other Essays*. Trans. Ben Brewster. New York and London: Monthly Review Press.

Bourdieu, Pierre (1987). *Distinction: A Social Critique of the Judgement of Taste*. Trans. Richard Nice. Cambridge, MA: Harvard University Press.

Brook, Peter (1968). *The Empty Space*. New York: Touchstone Books.

Cowan, Nelson (2001). "The Magical Number 4 in Short-term Memory: A Reconsideration of Mental Storage Capacity." *Behavioral and Brain Sciences*, 24, 87–114.

Dreyfus, Hubert (1997). "Intuitive, Deliberative, and Calculative Models of Expert Performance," in *Naturalistic Decision Making*. Ed. Caroline E. Zsambo and Gary Klein. Mahwah, NJ: Lawrence Erlbaum.

France 24 (2011). "Play in Paris Draws the Wrath of Far-Right Catholic Groups." Viewed February 17, 2018. http://www.france24.com/en/20111101-play-paris-draws-wrath-far-right-catholic-groups-christ-castellucci

Ingold, Tim (2015). *The Life of Lines*. London and New York: Routledge.

Lutterbie, John (2011). *Toward a General Theory of Acting*. New York: Palgrave McMillan.

Malabou, Catherine (2008). *What Should We Do with Our Brain?* New York: Fordham University Press.

Mauss, Marcel (2006). "Techniques of the Body," in *Techniques, Technology and Civilization*. Ed. Nathan Schlanger. New York and Oxford: Durkheim Press/ Berghahn Books.

Santayana, George (1905). *The Life of Reason: Reason in Common Sense*. New York: Scribner's.

Shockley, Gordon E. (2011). "Political Environment And Policy Change: The National Endowment For The Arts In The 1990s." *Journal of Arts Management, Law & Society*, 4 (41), 267–84.

Wise, Jason (2012). *Somm*. Documentary film.

Chapter 3

Barish, Jonas (1985). *The Antitheatrical Prejudice*. Berkeley: University of California Press.

Bodenhausen, Galen V., and Kurt Hugenberg (2015). "Attention, Perception, and Social Cognition," in *Social Cognition: The Basis of Human Interaction*. Ed. Fritz Strack and Jens Förster. New York: Psychology Press.

Brantley, Ben (2008). "Six Lies Ebb and Flow, Interconnected and Alone." *The New York Times*. November 16.

Brantley, Ben (2011). "Shakespeare Slept Here, Albeit Fitfully." *The New York Times*. April 13.

Brantley, Ben (2016). "'The Encounter' Is a High-Tech Head Trip through an Amazon Labyrinth." *The New York Times*. September 29. Viewed March 9, 2018. https://www.nytimes.com/2016/09/30/theater/the-encounter-review.html

Chabris, Christopher F., and Daniel J. Simons (2010). *The Invisible Gorilla: How Our Intuitions Deceive Us*. New York: Random House, Inc.

Chemero, Anthony (2003). "An Outline of a Theory of Affordances." *Ecological Psychology*, 15 (2), 181–95.

Clark, Andy (2016). *Surfing Uncertainty: Prediction, Action and the Embodied Mind*. Oxford: Oxford University Press.

Corbetta, Maurizio, and Gordon L. Shulman (2002). "Control of Goal-Directed and Stimulus-Driven Attention in the Brain." *Nature Reviews*, 3, 201–15.

Daily Motion (2012). *Einstein on the Beach: The Changing Image of Opera*. First aired by the Public Broadcasting Service (1986). Viewed March 6, 2018. https://www.dailymotion.com/video/xph2hm

Etchells, Tim (2008). *Forced Entertainment: On Making Performance*. Center for Performance Research. Viewed March 5, 2018. https://www.youtube.com/watch?v=m2fRfN5U7GA

Gardner, Lyn (2009). "We Are Waging a War." *The Guardian*. February 23. Viewed February 22, 2018. https://www.theguardian.com/stage/2009/feb/23/forced-entertainment-sheffield

Gilbert, Sophie (2012). "Theater Review: 'Red' at Arena Stage." *The Washintonian*. January 30. Viewed March 3, 2018. https://www.washingtonian.com/2012/01/30/theater-review-red-at-arena-stage/

Improv Everywhere (2008). *Frozen Grand Central*. Viewed March 4, 2018. https://improveverywhere.com/2008/01/31/frozen-grand-central/. For a video of the event: https://www.youtube.com/watch?v=jwMj3PJDxuo

Ingold, Tim (2015). *The Life of Lines*, London and New York: Routledge.

Jays, David (2002). "Joseph Svoboda." *The Guardian*. April 22. Viewed March 3, 2018. https://www.theguardian.com/news/2002/apr/22/guardianobituaries

Martin, Douglas (2002). "Josef Svoboda, 81, Designer of Hundreds of Productions." *The New York Times*. April 22. Viewed March 3, 2018. http://www.nytimes.com/2002/04/22/theater/josef-svoboda-81-stage-designer-for-hundreds-of-productions.html

Peterson, Steven E., and Michael I. Posner (2012). "The Attention System of the Human Brain: 20 Years After." *Annual Review of Neuroscience*, 35, 73–89.

Rainville-Tomson, Sydney (2012). "Little West 12th Night: Enlivening the Meatpacking District through Shakespeare." *Untapped Cities*. Viewed March 4, 2018. https://untappedcities.com/2012/07/18/little-west-12th-night-enlivening-the-meatpacking-district-through-shakespeare/

Riding, Alan (2000). "Theater; Peter Brook Prefers His 'Hamlet' Lean." *The New York Times*. December 10. Viewed March 4, 2018. http://www.nytimes.com/2000/12/10/theater/theater-peter-brook-prefers-his-hamlet-lean.html

Ryan, Marie-Laure. "Possible Worlds." *The Living Handbook of Narratology*. Viewed March 2, 2018. http://www.ihn.uni-hamburg.de/article/possible-words

Sheets-Johstone, Maxine (2000). "Kinetic Tactile-Kinesthetic Bodies: Ontogenetical Foundations of Apprenticeship Learning," from *Human Studies*, 23. Netherlands: Kluwer Academic Publishers.

Smith, David Woodruff (2013). "Phenomenology." *Stanford Encyclopedia of Philosophy*. Viewed February 22, 2018. https://plato.stanford.edu/entries/phenomenology/

Soloski, Alexis (2016). "The Encounter Review—Simon McBurney's Revolution in the Head." *The Guardian*. September 29. Viewed March 9, 2018. https://www.theguardian.com/stage/2016/sep/29/the-encounter-review-simon-mcburney-revolution-headphones

Stoppard, Tom (1967). *Rosencrantz and Guildenstern Are Dead*. New York: Grove Press.

Svoboda, Josef (2018). *J.S. – scenograf, o.p.s.* Viewed March 2, 2018. http://www.svoboda-scenograf.cz/en/

Chapter 4

Ashley, Richard (2012). "Music, Narrative, and Emotion." Conference on Memory, Emotion, and the Disciplines. Humanities Institute at Stony Brook University. March 22–23, 2012.

Barrault, Jean Louis (1951). *Reflections on Theatre*. London: Rockliff.

Beckett, Samuel (1984). *The Collected Shorter Plays of Samuel Beckett*. New York: Grove Press, Inc.

Brantley, Ben (2010). "Primary Colors and Abstract Appetites." *The New York Times*. June 27. Viewed April 11, 2018. https://www.nytimes.com/2010/04/02/theater/reviews/02red.html

Braun, Edward, ed. and trans. (1969). *Meyerhold on Theatre*. New York: Hill and Wang.

Cage, John (1961). *Silence: Lectures and Writings by John Cage*. Hanover, NH: Wesleyan University Press.

Clark, Andy (2015). *Surfing Uncertainty: Prediction, Action and the Embodied Mind*. Oxford: Oxford University Press.

Deleuze, Gilles (1992). *The Fold: Leibniz and the Baroque*. Trans. Tom Conley. Minneapolis: University of Minnesota Press.

Gallagher, Shaun and Dan Zahavi (2012). *The Phenomenological Mind*. Second edition. London and New York: Routledge.

Hodge, Francis, and Michael McClain (2010). *Play Directing: Analysis, Communication and Style*. Seventh edition. Waltham, MA: Focal Press.

Ingold, Tim (2015). *The Life of Lines*. London and New York: Routledge.

Johnson, Samuel M. (1976). "Two Lovers." *Einstein on the Beach: Libretto*. Viewed April 9, 2018. https://culturebox.francetvinfo.fr/sites/default/files/assets/documents/eob_text_for_libretto.pdf

Kroetz, Frans Xaver (1976). *Farmyard & Four Plays*. Trans. Michael Roloff and Jack Gelber. New York: Urizen Books.

Lakoff, George and Mark Johnson (1999). *Philosophy in the Flesh: The Embodied Mind and Its Challenge to Western Thought*. New York: Basic Books.

Levinas, Emmanuel (1969). *Totality and Infinity: An Essay on Exteriority*. Trans. Alphonso Lingis. Dordrecht and Norwell: Kluwer Academic Publishers.

Obenhaus, Mark (1985). *Einstein on the Beach: The Changing Image of Opera*. Viewed November 25, 2018. https://www.youtube.com/watch?v=HCIrNDgrQLE

Phùng, Thanh (2013). "Roland Barthes: Studium and Punctum." Museum of Education. Viewed March 11, 2018. https://educationmuseum.wordpress.com/2013/03/12/roland-barthes-studium-and-punctum/

Stanislavski, Constantin (1949). *Building a Character*. Trans. Elizabeth Reynolds Hapgood. New York: Theatre Arts Books.

Stein, Gertrude (1993), "Composition as Explanation," in *A Stein Reader*. Ed. Ulla E. Dydo. Evanston: Northwestern University Press.

Whitmann, Marc (2017). *Felt Time: The Science of How We Experience Time*. Trans. Eric Butler. Cambridge, MA: MIT Press.

Chapter 5

Althusser, Louis (1971). *Lenin and Philosophy and Other Essays*. Trans. Ben Brewster. New York and London: Monthly Review Press.

Ashley, Richard (2012). "Music, Narrative, and Emotion." Conference on Memory, Emotion, and the Disciplines. Humanities Institute at Stony Brook University. March 22–23, 2012.

Beckett, Samuel (1984). *The Collected Shorter Plays of Samuel Beckett*. New York: Grove Press, Inc.

Beckett, Samuel, and Billie Whitelaw (1973). *Not I*. Accessed November 24, 2018. https://www.youtube.com/watch?v=M4LDwfKxr-M

Blau, Herbert (1987). *Performance Theory: Modern Drama and Postmodern Theater*. National Endowment for the Humanities Summer Seminar, University of Wisconsin-Milwaukee.

Brown, Margaret Wise (1942). *The Runaway Bunny*. New York: Harper.

Cage, John (1961). *Silence: Lectures and Writings by John Cage*. Hanover, NH: Wesleyan University Press.

Cook, Amy (2018). *Building Character: The Art and Science of Casting*. Ann Arbor: University of Michigan Press.

Csikszentmihályi, Mihaly (1990). *Flow: The Psychology of Optimal Experience*. New York: Harper and Row.

Everett, Daniel (2017). "Chomsky, Wolfe and Me." *AEON*. Viewed October 28, 2018. https://aeon.co/essays/why-language-is-not-everything-that-noam-chomsky-said-it-is

Fauconnier, Gilles, and Mark Turner (2002). *The Way We Think: Conceptual Blending and the Mind's Hidden Complexities.* New York: Basic Books.

Fornes, Maria Irene (1986). *Plays.* New York: PAJ Publications.

Gallese, Vittorio (2008). "Mirror Neurons and the Social Nature of Language: The Neural Exploitation Hypothesis." *Social Neuroscience,* 3 (3–4), 317–33.

Gendlin, Eugene T. (1962). *Experiencing and the Creation of Meaning a Philosophical and Psychological Approach to the Subjective.* Evanston: Northwestern University Press.

Gendlin, Eugene T. (2004). "The New Phenomenology of Carrying Forward." *Continental Philosophy Review,* 37, 127–51.

Greicius, Michael D., Ben Krasnow, Allan L. Reiss, and Vinod Menon (2003). "Functional Connectivity in the Resting Brain: A Network Analysis of the Default Mode Hypothesis." *Proceedings of the National Academy of Sciences of the United States of America,* 100 (1), 253–8.

Hodge, Francis, and Michael McClain (2015). *Play Directing: Analysis, Communication and Style.* Seventh edition. Waltham, MA: Focal Press.

Ingold, Tim (2015). *The Life of Lines.* London and New York: Routledge.

Kroetz, Frans Xaver (1976). *Farmyard & Four Plays.* Trans. Michael Roloff and Jack Gelber. New York: Urizen Books.

Lakoff, George, and Mark Johnson (1999). *Philosophy in the Flesh: The Embodied Mind and Its Challenge to Western Thought.* New York: Basic Books.

Phùng, Thanh (2013). "Roland Barthes: Studium and Punctum." Museum of Education. Viewed March 11, 2018. https://educationmuseum.wordpress.com/2013/03/12/roland-barthes-studium-and-punctum/

Pinter, Harold (2014). *No Man's Land.* New York: Grove Press.

Reber, Rolf (2014). "Processing Fluency, Aesthetic Pleasure, and Culturally Shared Taste," in *Aesthetic Science: Connecting Minds, Brains, and Experience.* Ed. Arthur P. Shimamura and Stephen E. Palmer. Oxford: Oxford University Press.

Skipper, Jeremy I., Joseph T. Devin, and Daniel R. Lametti (2017). "The Hearing Ear Is Always Found Close to the Speaking Tongue: Review of the Role of the Motor System in Speech Perception." *Brain and Language,* 164, 77–105.

Turney, Peter (2009). "Criticisms of Lakoff's Theory of Metaphor." *Apperceptual.* Viewed April 30, 2018. http://blog.apperceptual.com/criticisms-of-lakoff-s-theory-of-metaphor

Chapter 6

Baitz, Jon Robin (2011). *Other Desert Cities.* Dramatist Play Service. Viewed June 12, 2018. https://www.dramatists.com/cgi-bin/db/single.asp?key=4386

Brecht, Bertolt (1964). *Brecht on Theatre*. Trans. John Willet. New York: Hill and Wang.

Faruqee, Anoka (2009). "The Kantian Sublime: Why Care?" "Why Theory," Cal Arts Exhibition Catalogue. Viewed May 17, 2018. http://anokafaruqee.com/the-kantian-sublime-why-care/

Gallese, Vittorio, and David Freedberg (2007). "Motion, Emotion and Empathy in Esthetic Experience." *TRENDS in Cognitive Science*, 11(5), 197–203.

Jauss, Hans Robert (1982). *Toward an Aesthetic of Reception*. Trans. Timothy Bahti. Minneapolis: University of Minnesota Press.

Lutterbie, John (2011). *Toward a General Theory of Acting*. New York: Palgrave McMillan.

Proust, Marcel (1981). *Remembrance of Things Past*. Trans. C. K. Scott Moncreiff and Terence Kilmartin. New York: Vintage Books.

Ramachandran, V. S. (2011). *The Tell-Tale Brain: A Neuroscientists Quest for What Makes Us Human*. New York and London: W. W. Norton and Company.

Shklovsky, Victor (1988). "Art as Technique," in *Modern Criticism and Theory: A Reader*. Ed. David Lodge. London and New York: Longman, 16–30.

Smithson, Robert (1979–80). *Mirrors and Shelly Sand*. Dallas Museum of Art. Accessed November 24, 2018. https://collections.dma.org/artwork/5324617

Epilogue

Beadle-Brown, Julie, David Wilkinson, Lisa Richardson, Nicola Shaughnessy, Melissa Trimingham, Jennifer Leigh, Beckie Whelton, and Julian Himmerich (2018). "Imagining Autism: Feasibility of a Drama-based Intervention on the Social, Communicative and Imaginative Behaviour of Children with Autism." *Autism*, 22 (8), 915–27. Viewed June 17, 2018. http://journals.sagepub.com/doi/abs/10.1177/1362361317710797#articleCitationDownloadContainer

Malabou, Catherine (2008). *What Should We Do with Our Brain?* New York: Fordham University Press.

Merleau-Ponty, Maurice (1968). *The Visible and the Invisible*. Ed. Claude LeFort. Trans. Alphonso Lingis. Evanston: Northwestern University Press.

Mudugno, Nicola, Sara Iaconelli, Mariagrazia Fiorilli, Francesco Lena, Imogen Kusch, and Giovanni Mirabella (2010). "Active Theater as a Complementary Therapy for Parkinson's Disease Rehabilitation: A Pilot Study." *The Scientific World JOURNAL*, 10, 2301–13.

Parsons, Zack (2013). "The Full Version of Roy Batty's 'Tears in Rain' Speech." Viewed April 5, 2019. https://www.somethingawful.com/news/blade-runner-speech/1/

Further Reading

Blair, Rhonda (2008). *The Actor, Image, and Action: Acting and Cognitive Neuroscience*. London and New York: Routledge.
Rhonda Blair's book is one of the first to apply cognitive science to the art of acting. She provides an excellent introduction to the value of utilizing science in the study of theatre in general and to acting in particular. The work is grounded on the acting theories of Stanislavski, which are elucidated by scientific concepts, such as mirror neurons, cognitive linguistics, the study of metaphor, imagination, and consciousness. Case studies show how the theory is applied in productions directed by the author.

Cook, Amy (2010). *Shakespearean Neuroplay: Reinvigorating the Study of Dramaic Texts and Performance through Cognitive Science*. New York: Palgrave Macmillan.
Cognitive integration (blending) is applied to Shakespeare's *Hamlet*. Cook uses the "mirror held up to nature" as a key for understanding how the play means. In an early chapter, cognitive integration is defined in depth. She makes the larger and exciting claim that every generation reads the play through the lens of their contemporary society, but we use the cognitive functions that were used by Shakespeare's audience.

Cook, Amy (2018). *Building Character: The Art and Science of Casting*. Ann Arbor: University of Michigan Press.
Amy Cook applies cognitive integration (blending) as the theoretical underpinnings for looking at the way casting affects the meaning that audiences derive from a production. Shakespearean productions are at the core of her research, but her case studies cover a range of modern and post-modern productions. The work focuses on categorization and its relation to identity, arguing that our actions can change how we are perceived as performers on and off the stage.

Crane, Mary Thomas (2000). *Shakespeare's Brain: Reading with Cognitive Theory.* Princeton: Princeton University Press.

Mary Thomas Crane considers the brain as a site where body and culture meet to form the subject and its expression in language. Taking Shakespeare as her case study, she boldly demonstrates the explanatory power of cognitive theory—a theory which argues that language is produced by a reciprocal interaction of body and environment, brain and culture, and which refocuses attention on the role of the author in the making of meaning. Crane reveals in Shakespeare's texts a web of structures and categories through which meaning is created.

Falletti, Clelia, Gabriele Sofia, and Victor Janco (2016). *Theatre and Cognitive Neuroscience.* London, Oxford, New York, and New Delhi: Bloomsbury.

This edited collection focuses on actor/audience relationship and is divided into four sections with essays written by an international set of authors. The first section looks at spatial dynamics of the interaction between performer and spectator. The second and third focus on the viewer and actor respectively. The fourth looks at applied theatre and how theatre can serve as an intervention with patients with cognitive disabilities. The volume is notable for its combination of theatre practitioners and cognitive scientists from a number of disciplines.

Kemp, Rick (2012). *Embodied Acting: What Neuroscience Tells Us about Performance.* Abingdon, UK, and New York: Routledge.

Kemp combines research in the cognitive sciences with practice based on his work as an actor and director. Exploring many of the ideas in this book, such as metaphor and conceptual blending, he includes acting exercises that allow the reader to put the theories into practice. It is particularly useful for enhancing the connection between the physical and the text.

Kemp, Rick, and Bruce McConachie (2019). *The Routledge Companion to Theatre, Performance and Cognitive Science.* Abingdon, New York. Routledge.

A wide-ranging survey in which leading authors in the field address four broad questions: How can performances in theatre, dance, and other media achieve more emotional and social impact? How can we become more adept teachers and learners of performance both within and outside of classrooms? What can the cognitive sciences reveal about drama and human nature in general? How can Knowledge Transfer from a synthesis of science and performance assist professionals such as nurses, care-givers, therapists, and emergency workers in their jobs?

McConachie, Bruce (2008). *Engaging Audience: A Cognitive Approach to Spectating in the Theatre*. New York: Palgrave Macmillan.
One of the first books to discuss audience response in the theatre from a cognitive science perspective. It weaves together theatre history and the spectator's experience, using a blend of cognitive science and Lakoff and Johnson's embodied metaphors to ground the discussion.

Mermikides, Alex, and Gianna Bouchard (2016). *Performance and the Medical Body*. London, Oxford, New York, and New Delhi: Bloomsbury.
This volume brings together essays by artists, scientists, and scholars to explore the relationship between performance and the medical body. The three sections investigate the institution of medicine, patients, and the body. The editors write, "Emerging from a range of perspectives and utilizing different concepts of 'performance', these essays offer insights into [the] interface of [medicine and performance]."

Paavolainen, Teemu (2012). *Theatre/Ecology/Cognition: Theorizing Performer-Object Interaction in Grotowski, Kantor, and Meyerhold*. New York: Palgrave Macmillan.
Paavolainen draws on cognitive theories of affordances and image schemas to engage major works by Meyerhold, Grotowski, and Kantor. He argues that the ecologies of the theatre in combination with 4-E cognition challenge the subject/object dichotomy implicitly in thinking about theatre. His insights extend beyond the limit of his case studies, offering a new way deriving meaning in the theatre as audience and practitioner.

Page, Kevin (2018). *Advanced Consciousness Training for Actors: Meditation Training for Actors*. New York and London: Routledge.
A practical and theoretical book, Page's work "explores theories and techniques for deepening the individual actor's capacity to concentrate and focus attention ... These practices utilize consciousness expanding 'technologies' derived from both Eastern and Western traditions of meditation and mindfulness and neuroscience. This book reviews the scientific literature of consciousness studies and mindfulness research to discover techniques ... foundational skills of the performing artist in any medium."

Rokotniz, Naomi (2011). *Trusting Performance: A Cognitive Approach to Embodiment in Drama*. New York: Palgrave Macmillan.
The cognitive basis for trust is examined in four plays spanning the modern era. The four chapters explore the author/performer/spectator networks of meaning making. Concepts of empathy and mirroring systems provide the background for the exploration of the vicissitudes of putting our trust in other people. Trust is seen as something that occurs not in the brain but in our embodied

interactions with each other. Indeed, the networks we develop with others, for good or ill, are founded in our recognition of visceral affectivity of that engagement.

Tribble, Evelyn B. (2011). *Cognition in the Globe: Attention and Memory in Shakespeare's Theatre*. New York: Palgrave Macmillan.
Evelyn Tribble explores the prodigious demands on the memory of actors in Shakespeare's Globe Theatre. A wide-ranging book, she brings together questions of memory, attention, perception, and affordances in a study that also includes the makeup of the company, the distribution of parts, and actor/ audience interaction. The result is a dynamic model of the material conditions facing his performers.

Index